Existence Theorems for Ordinary Differential Equations

Existence Theorems for Ordinary Differential Equations

by FRANCIS J. MURRAY, Ph.D.
Professor Mathematics, Duke University

and

KENNETH S. MILLER, Ph.D.
Riverside Research Institute
and
Adjunct Professor of Mathematics, Fordham University

ROBERT E. KRIEGER PUBLISHING COMPANY
Huntington, New York
1976

Library of Congress Cataloging in Publication Data

Murray, Francis Joseph, 1911–
Existence theorems for ordinary differential equations.

Reprint of the ed. published by New York University Press, New York.
1. Differential equations. 2. Existence theorems. I. Miller, Kenneth S., joint author.
II. Title.
QA371.M985 1975 515'.352 75-12685
ISBN 0-88275-320-7

ORIGINAL EDITION 1954
REPRINTED 1976

Printed and Published by
ROBERT E. KRIEGER PUBLISHING COMPANY
645 NEW YORK AVENUE
HUNTINGTON, NEW YORK 11743

Printed in the United States of America

Contents

Introduction

The modern physical sciences developed from natural philosophy as men realized that it was possible to treat motion and time rates in general by mathematical methods. These methods in the case of Leibnitz and his followers led to "differential equations" as a method for describing phenomena while the "fluxions" of Newton indicated derivative relations. The early workers knew that they were on the right track, that is, that differential or derivative equations did actually correspond to physical phenomena but logical difficulties were all too apparent in the original crude approach. These difficulties were seized upon by critics such as Berkeley in an effort to discredit the whole movement.

Precisely these difficulties were responsible for the development of analysis. Modern real function theory is the answer to the difficulties of "infinitesimals" and the formal use of series raised questions that led to the essential developments in the complex theory. The Cauchy theory was shaped for the needs of differential equations and those questions which we now consider the most basic were formulated by him and the first answers given.

Thus analysis in its "nascent" stage was identical with the theory of differential equations. During the nineteenth century the theory of differential equations continued to play a major role. The distinction between ordinary and partial differential equations is as old as Euler and the corresponding theories for the analytic case tended to have independent developments. Both are almost necessarily theories for functions of a number of variables but in ordinary differential equations all but one variable appear "parametrically," that is, no derivatives with respect to these variables appear. The analytic theory was continuously expanded and simplified. The early majorant methods of Cauchy eventually were replaced by the Picard iterant procedures which still yield the required analytic dependence of the solutions on the variables.

The precise definition of a real number and the precise formulation of the theory of continuous functions (developments stemming from the efforts of Fourier to solve the partial differential equations of heat flow) demanded a reformulation of the theory of differential equations in which hypotheses on the differential equations would be restricted to continuity or the existence of derivatives of some specified small order and the solutions sought were required merely to have continuous derivatives of the order which actually appeared in the equation. The practical value of such a development is not necessarily apparent; but when the theory was available, the structure indicated showed the essential aspects of the proofs and permitted simplifications even in the analytic case.

However, general existence theorems are either for the linear case or for a small neighborhood around a specified point or for solutions which surround a given known solution. These limitations have led in the present century to efforts to obtain theories "in the large" such as those of Poincaré or the theories applicable to geodesics. Many special differential equations have been studied and there have been numerous investigations of periodic and asymptotic properties of solutions and stability.[1]

The objective in the present book is not to go as far as this latter modern work. We assume a knowledge of basic real function theory (and for certain specialized results only, of elementary functions of a complex variable) and establish the fundamental existence theorems for ordinary differential equations which are the culmination of the nineteenth-century development. We do not consider the elementary methods for solving certain special differential equations nor the more advanced specialized topics.[2] By restricting ourselves in this fashion we hope to obtain a logically coherent discussion which has educational value for the student in a certain stage of his mathematical development.

Certain developments which have occurred since 1946 have given these existence theorems and the theorems on parameters a new importance in applied mathematics. Until recently the possibilities

[1] Cf. R. Bellman, "A survey of the theory of the boundedness, stability, and asymptotic behavior of solutions of linear and non-linear differential and difference equations," Jan. 1949, Office of Naval Research, Department of the Navy, Washington, D. C.

[2] Cf. E. L. Ince, "Ordinary differential equations," Longmans, Green and Co., Ltd., 1927.

of finding solutions in practical form were very limited. The solutions of certain differential equations have been tabulated by long tedious computation processes. For a limited region, the Taylor series solutions can be used, but many interesting and important properties of solutions are not obtainable in a practical way from series. Except in the case of differential equations with constant coefficients, these were the only possibilities.

However, in the post-war period, electronic differential analyzers have become available for readily producing graphical representations of solutions. The dependence of the solution on parameters can be exhibited by taking many solutions with different values of the parameters. Thus the general long range properties of solutions are available and their existence can be checked by digital procedures.

Much has been done with these machines effectively by "analog methods," that is, reasoning based on direct analogies between the original system to be studied and an electronic system constructed from differential analyzer components. There are, however, always crudities in these analogies. In the simpler cases one may be justified in ignoring the discrepancies between the original system and the model, but a thorough understanding of the situation requires an analysis of the differential equations which govern both phenomena.

These questions have been investigated by the authors,[1] and it is surprising how effectively this classical theory can be used to specify the relation between the two systems. If the order of the two systems is the same it is in general immediately applicable. If the order is raised, a basic extension must be made because non-analyticities appear; but the classical theory is still necessary for this extension. There are many arguments based on "linearizing" the problem which can be replaced by exact discussions using the classical theory of differential equations.

The long range importance of the use of electronic differential analyzers cannot be overestimated. They greatly extend the cases to which scientific analysis is possible. Indeed, one was previously limited in practice to certain cases governed by linear differential

[1] K. S. Miller and F. J. Murray, *A mathematical basis for an error analysis of differential analyzers*, Jn. of Math. and Phys., *32*, Nos. 2–3, 136–163, July—October 1953.

equations with constant coefficients. The differential analyzer permits one quickly to explore situations in a manner equivalent to the most refined mathematical estimates of growth and long range behavior. Practically, they eliminate an enormous amount of expensive experimentation by narrowing the range within which the desired behavior occurs. In a design it is frequently found that the importance of different factors varies tremendously. The differential analyzer permits one to eliminate quickly minor factors and concentrate on the important ones. This is a mathematical job which would be quite difficult by non-machine methods.

But clearly the use of these devices is just the initial step in a development which should eventually permit the full use of scientific knowledge for technical purposes. Improved accuracy should result in eliminating even more of the required experimentation, and the introduction and use of auxiliary digital computation would be a further advance. A better technology in turn will indicate where scientific knowledge can itself be improved. For the limits of such a technology will correspond to limits of scientific knowledge.

It is believed that the mathematical theories here presented will continue to be of importance in the above mentioned development. When analog devices are used without a proper mathematical basis, the user appears to be groping in the dark and indeed much of the "experience" acquired under these circumstances has precisely the unpleasant character associated with barking one's shins on an unseen obstacle. The mathematical basis is precisely the illumination needed to map out the general situation.

January 1954

F. J. M.
K. S. M.

CHAPTER 1

The Fundamental Existence Theorems

1. The basic existence theorem

1.1 Let y be an unknown function of x. Frequently our information on y is its initial value y_0 for $x = x_0$ and its rate of change dy/dx. If our information on dy/dx is that dy/dx is a continuous function f(x), then we know that there is a unique y, which may be obtained by integration. If, on the other hand, as often happens, our information specifies dy/dx in terms of y as well as x, that is, $dy/dx = f(y, x)$[1] then even if f depends continuously on y and x the situation is not clear. For this equation on dy/dx does not yield prima facie evidence of the existence of y, since to evaluate f, y must be known, that is, y must exist.

Clearly then the existence of y must be established by some building-up process. Since $\frac{dy}{dx} = \lim_{\Delta x \to 0} \frac{\Delta y}{\Delta x}$ there is an interval $(x_0, x_0 + h)$ such that y is well approximated by $y_0 + f(y_0, x_0) \cdot (x - x_0)$. If $x_1 = x_0 + h$ and $y_1 = y_0 + f(y_0, x_0)\, h$ we can take a second interval $(x_1, x_1 + h)$ in which y is approximated by $y_1 + f(y_1, x_1)\,(x - x_1)$. Continuation of this process will yield a polygonal line with slope in the form $f(y_1, x_1)$ except at the vertices where in general there is not a unique slope. Intuitively, we would expect that if h is taken smaller and smaller the corresponding polygonal lines would converge to a solution y of the differential equation $dy/dx = f(y, x)$.

However, the difficulties of the situation appear as soon as we try to make this procedure precise. Question one is, how do we know these lines will converge? Secondly, if they do, will the resulting

[1] We have written f(y, x) instead of perhaps the more conventional f(x, y). However, the notation f(y, x) extends more naturally to the case of n dependent variables, viz.: $f(y_1, y_2, \ldots, y_n, x)$.

function have a derivative? (The approximants in general will fail to have derivatives at more and more points as we take h smaller and smaller.) Thirdly, if the limiting function exists and has a derivative, will it satisfy the given differential equation?

Relative to the first question one can show that if f is continuous one can choose a sequence of h's approaching zero such that the polygonal lines will converge for some x interval with lower end point x_0. However, in general, the size of this interval may be very small and has to be investigated in each individual case.

For questions two and three one has available certain technical devices. One transforms the problem from one in differential equations to an equivalent integral equation

$$y = y_0 + \int_{x_0}^{x} f(y, t)\, dt.$$

Then if our approximants converge *uniformly* we obtain the desired results.

Precisely formulated,[1] the fundamental existence theorem reads as follows:

Theorem 1. Let f(y, x) be a real valued function of the two real variables y, x defined and continuous[2] on an open region $\mathfrak{A}$ of two dimensional euclidean space. Then for every point (y_0, x_0) of $\mathfrak{A}$ we can find a $b > 0$ and a function $\varphi(x)$ which has a continuous first derivative in the neighborhood $\mathfrak{N}$, $|x - x_0| \leq b$ such that

$$\frac{d\varphi}{dx} = f(\varphi(x), x)$$

and $\varphi(x_0) = y_0$ in this neighborhood $\mathfrak{N}$.

We shall prove Theorem 1 in Section 1.8. Note that the theorem has been stated for the case of *one* unknown function y. This will be generalized in later sections to n functions, implicit forms and higher derivatives. However, in order to appreciate clearly the ideas underlying the theory, we shall carry through the initial discussion for the one unknown function y.

[1] Theorem 1 was proved essentially in the form we have given above by G. Peano in the latter part of the nineteenth century. The theorem and its generalizations are frequently referred to as the "Peano theorem."

[2] When we say f(y, x) is continuous in y and x we shall always mean *joint continuity*, that is, given an $\varepsilon > 0$ there exists a $\delta > 0$ such that $|f(y, x) - f(y', x')| < \varepsilon$ whenever $|x - x'| < \delta$ and $|y - y'| < \delta$.

It is important to note that Theorem 1 is an *existence theorem* and not a *uniqueness theorem*. Under the hypothesis of the theorem, cases exist in which there exist two or more distinct functions satisfying the conclusions of the theorem. Examples will be given later (cf. Section 1.1 of Chapter 3) to illustrate this phenomenon.

1.2 We shall unfold the theory as a logical entity. As we attempt to prove it, it will appear that certain auxiliary lemmas are necessary. These we shall state and prove only after we have seen the necessity for such a digression.

We are concerned with the differential equation

$$\frac{dy}{dx} = f(y, x) \tag{1}$$

where $f(y, x)$ satisfies the conditions of Theorem 1. Suppose $b > 0$ is assigned and let $\mathfrak{N}$ be the set of x's with $|x - x_0| \leq b$. Suppose further that there exists a function $y(x)$ defined for $x \varepsilon \mathfrak{N}$ which satisfies Equation (1) and at $x = x_0$ assumes the value y_0. Of necessity, dy/dx must exist and hence $y(x)$ is continuous. Since f is continuous, $f(y(x), x)$ is a continuous function of x in the neighborhood $\mathfrak{N}$ of $x = x_0$. Hence, since $dy/dx = f(y(x), x)$, dy/dx is continuous, and consequently

$$y - y_0 = \int_{x_0}^{x} \frac{dy}{dt}\, dt = \int_{x_0}^{x} f(y(t), t)\, dt$$

for $x \varepsilon \mathfrak{N}$. Equation (1) therefore implies

$$y = y_0 + \int_{x_0}^{x} f(y(t), t)\, dt. \tag{2}$$

On the other hand, a function y that satisfies Equation (2) must be a continuous function of x. For, if y satisfies Equation (2) it implies that $f(y(t), t)$ is Riemann integrable in $\mathfrak{N}$ and hence $y_0 + \int_{x_0}^{x} f(y(t), t)\, dt$ is continuous. But this expression is precisely y. Now since y is continuous, and f is continuous by hypothesis, $f(y(t), t)$ is continuous for $t \varepsilon \mathfrak{N}$. Thus the derivative of the integral on the right hand side of Equation (2) exists and equals the integrand. Consequently, differentiating Equation (2) will yield Equation (1).

We have thus proved the following lemmas:

Lemma 1. If there exists a function $\varphi(x)$ defined for $x \varepsilon \mathfrak{N}$ which satisfies Equation (1), then $d\varphi/dx$ exists and is continuous.

Lemma 2. The solutions of Equation (1) which at x_0 equal y_0 are identical with the solutions of Equation (2).

The study of the differential equation, Equation (1), has thus been reduced to the study of the integral equation, Equation (2). As stated above, this will permit us eventually to answer questions two and three.

1.3 In the statement of Theorem 1, a neighborhood $\mathfrak{N}$ of x_0, namely $|x - x_0| \leq b$ appeared in the conclusion. The construction of this neighborhood as well as a discussion of its significance will now be given.

Since $\mathfrak{A}$ is an open region, there exists a square of side 2a with center at (x_0, y_0) such that the entire square including its boundaries is in the region $\mathfrak{A}$. This square, consisting of the points (y, x)

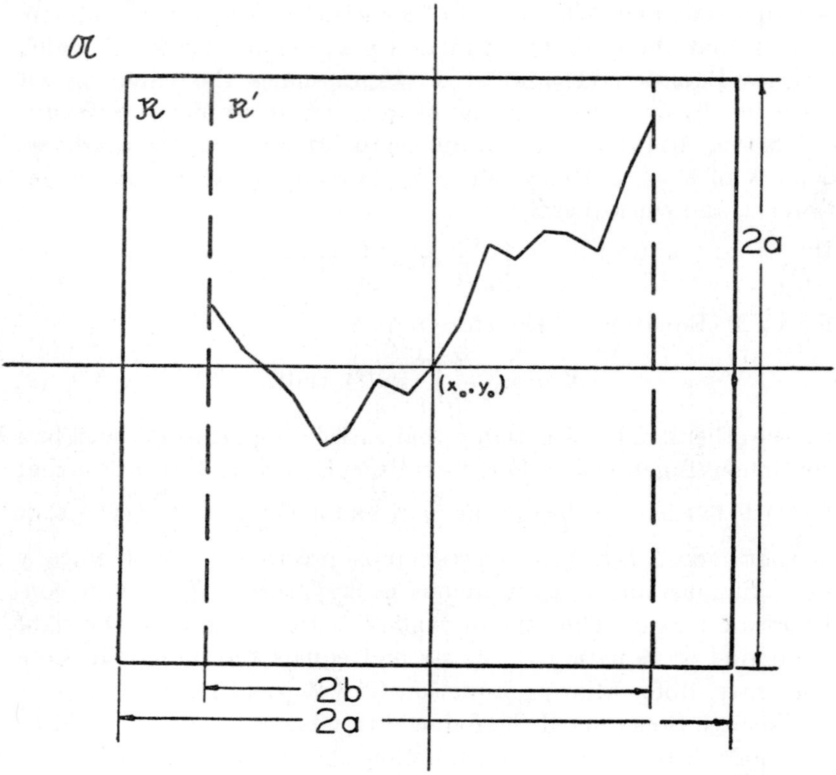

FIGURE 1

such that $|x - x_0| \leq a$, $|y - y_0| \leq a$ will be called $\mathfrak{R}$. (Cf. Figure 1.) Since $\mathfrak{R}$ is a closed point set interior to $\mathfrak{A}$ and $f(y, x)$ is continuous on $\mathfrak{A}$, f is bounded on $\mathfrak{R}$. That is, there exists a constant M such that

$$|f(y, x)| \leq M \quad \text{for} \quad (y, x) \in \mathfrak{R}.$$

Now let

$$b = \min\left(a, \frac{a}{M}\right)$$

and consider the set of points $|x - x_0| \leq b$, $|y - y_0| \leq a$. We shall call this closed rectangular subregion of $\mathfrak{R}$, $\mathfrak{R}'$. Clearly,

(i) $f(y, x)$ is continuous in $\mathfrak{R}'$ and $|f(y, x)| \leq M$ in $\mathfrak{R}'$.

(ii) If $|x - x_0| \leq b$ and $|y - y_0| \leq M|x - x_0|$, then $(y, x) \in \mathfrak{R}'$.

To prove (i) we merely note that $\mathfrak{R}' \leq \mathfrak{R} < \mathfrak{A}$. To demonstrate (ii), we note that

$$|y - y_0| \leq M|x - x_0| \leq Mb \leq M\left(\frac{a}{M}\right) = a.$$

As will be seen later, the reason for introducing b is to make (ii) a true statement.

Our next task is to construct a polygonal line function which is intended to approximate a solution to Equation (1). Let $h > 0$ be given. We shall suppose that h is small relative to b. Furthermore, it is convenient to assume that b is an integral multiple of h, that is,

$$b = ph$$

where p is a positive integer.

We define inductively a sequence of pairs of points $y_1, x_1; \ldots; y_p, x_p = x_0 + b$ by means of the relations

$$y_j = y_{j-1} + f(y_{j-1}, x_{j-1})\,h$$
$$x_j = x_{j-1} + h.$$

In order to make this definition valid we must prove that the $f(y_j, x_j)$ exist. By hypothesis it exists for $j = 0$, and furthermore $|f(y_0, x_0)| \leq M$. Now suppose $f(y_0, x_0), \ldots, f(y_{j-1}, x_{j-1})$ exist and are less than M in absolute value. Then since

$$y_j - y_0 = \sum_{k=1}^{j} (y_k - y_{k-1}) = \sum_{k=1}^{j} f(y_{k-1}, x_{k-1})\,h$$

we have

$$|y_j - y_0| \leq \sum_{k=1}^{j} |f(y_{k-1}, x_{k-1})|\,h \leq M\,jh = M|x_j - x_0|.$$

But by (ii) above this means that (y_j, x_j) is in $\mathfrak{R}'$ and thus $f(y_j, x_j)$ exists, and by (i), $|f(y_j, x_j)| \leq M$.

We now define a function $y(x, h)$ by the equation

$$y(x, h) = y_{j-1} + f(y_{j-1}, x_{j-1})(x - x_{j-1}) \tag{3}$$

for

$$0 < x - x_{j-1} \leq h$$

and

$$j = 1, 2, \ldots, p.$$

The quantity $y(x_0, h)$ is defined as

$$y(x_0, h) = y_0.$$

Note that the x_0 and y_0 are the initial point of Theorem 1.

The function $y(x, h)$ has thus been defined for all x in the interval $x_0 \leq x \leq x_0 + b$. It is a continuous function of x consisting of a finite number of straight line segments. A similar definition of $y(x, h)$ can be given for the interval $x_0 - b \leq x < x_0$ by letting $y(x, h)$ equal $y_{j+1} + f(y_{j+1}, x_{j+1})(x - x_{j+1})$ for $-h \leq x - x_{j+1} < 0$ where $y_j = y(x_j, h)$, $x_j = x_0 + jh$ and $j = -1, -2, \ldots, -p$.

Now in the region $x_0 \leq x \leq x_0 + b$ define the function $F(x, h)$ by the equation

$$F(x, h) = f(y_{j-1}, x_{j-1}) \tag{4}$$

for $0 < x - x_{j-1} \leq h$, $j = 1, 2, \ldots, p$ and in the region $x_0 - b \leq x < x_0$ let $F(x, h)$ be defined by

$$F(x, h) = f(y_{j+1}, x_{j+1}) \tag{5}$$

for $-h \leq x - x_{j+1} < 0$, $j = -1, -2, \ldots, -p$.

From Equation (3) we conclude that

$$y(x, h) = y_0 + \sum_{k=0}^{j-2} f(y_k, x_k)\, h + f(y_{j-1}, x_{j-1})(x - x_{j-1}),$$

$$j > 1, \quad 0 < x - x_{j-1} \leq h.$$

Also

$$y(x, h) = y_0 - \sum_{k=0}^{j+2} f(y_k, x_k)\, h + f(y_{j+1}, x_{j+1})(x - x_{j+1}),$$

$$j < -1, \quad -h \leq x - x_{j+1} < 0$$

while

$$y(x, h) = y_0 + f(y_0, x_0)(x - x_0), \quad -h \leq x - x_0 \leq h.$$

From these equations and Equations (4) and (5) we see that

$$y(x, h) = y_0 + \int_{x_0}^{x} F(t, h)\, dt, \qquad x_0 - b \leq x \leq x_0 + b. \tag{6}$$

As mentioned above, $y(x, h)$ is a continuous function of x. Also, we have shown above that $(y_j, x_j) \in \mathfrak{R}'$ and consequently

$$| F(t, h) | = | f(y_j, x_j) | \leq M. \tag{7}$$

But from this result and Equation (6),

$$| y(x, h) - y_0 | \leq M \, | x - x_0 |$$

which by virtue of (ii) implies

$$(y(x, h), x) \in \mathfrak{R}'$$

for all x, $| x - x_0 | \leq b$.

Since $F(t, h)$ is a step function with a finite number of jumps, Equation (6) defines a continuous function of x and we may write

$$y(x, h) = y_0 + \int_{x_0}^{x} f(y(t, h), t)\, dt + u(x, h) \tag{8}$$

where, by definition,

$$u(x, h) = \int_{x_0}^{x} [F(t, h) - f(y(t, h), t)]\, dt. \tag{9}$$

1.4 We recall that since $\mathfrak{R}$ is compact and $f(y, x)$ is continuous on $\mathfrak{R}$, f is uniformly continuous. Hence given an $\varepsilon > 0$ there exists a $\delta > 0$ such that $| f(y', x') - f(y'', x'') | < \varepsilon$ for all pairs of points (x', y'), (x'', y'') in $\mathfrak{R}$ with $| x' - x'' | < \delta$ and $|y' - y'' | < \delta$. We shall refer to this $\delta = \delta(\varepsilon)$ as the "uniform δ for ε" — referring of course to the function $f(y, x)$ in $\mathfrak{R}$.

Lemma 3. Let $\varepsilon > 0$ be assigned. Let $\delta > 0$ be the uniform δ for ε/b and let $h_0 = \min(\delta, \delta/M)$. Then if $h < h_0$ and $| x - x_0 | \leq b$,

$$| u(x, h) | \leq \varepsilon.$$

Proof. Suppose $x \geq x_0$ and $0 < x - x_{j-1} \leq h$. From Equation (3)

$$| y(x, h) - y_{j-1} | = | f(y_{j-1}, x_{j-1}) | \cdot | x - x_{j-1} | \leq Mh.$$

But,

$$| x - x_{j-1} | \leq h < h_0 \leq \delta$$

and

$$| y - y_{j-1} | \leq Mh < Mh_0 \leq \delta$$

by the hypothesis of the lemma. We have immediately

$$\frac{\varepsilon}{b} > | f(y, x) - f(y_{j-1}, x_{j-1}) | = | f(y, x) - F(x, h) |$$

and

$$| u(x, h) | \leq \left| \int_{x_0}^{x} | F(t, h) - f(y(t, h), t) | \, dt \right| < \frac{\varepsilon}{b} | x - x_0 | \leq \varepsilon.$$

A similar argument holds for $x_0 - b \leq x < x_0$.

1.5 In Lemma 4 we shall investigate the polygonal functions as h approaches zero.

Lemma 4. Let $h_1, h_2, \ldots$ be a sequence of positive numbers such that $\lim_{n\to\infty} h_n = 0$. Suppose that for $| x - x_0 | \leq b$, $y(x, h_n)$ converges uniformly to a function $y(x)$. Then for $|x_0 - x | \leq b$,

$$y(x) = y_0 + \int_{x_0}^{x} f(y(t), t) \, dt.$$

Proof. Consider Equation (8) with h replaced by h_n. Lemma 3 implies that

$$u(x, h_n) \to 0 \tag{10}$$

uniformly for any $x \varepsilon \mathfrak{R}$. Also, since the $y(x, h_n)$ converge uniformly to $y(x)$ by hypothesis, $y(x)$ is continuous. We assert then that

$$\lim_{n\to\infty} f(y(x, h_n), x) = f(y(x), x)$$

uniformly in x. To prove this, let $\varepsilon > 0$ be assigned. Then since $f(y, x)$ is uniformly continuous for $(y, x) \varepsilon \mathfrak{R}$ there exists a uniform $\delta > 0$ such that

$$|f(y(x, h_n), x) - f(y(x), x) | < \varepsilon$$

whenever $| y(x, h_n) - y(x) | < \delta$ and (trivially) $| x - x | < \delta$. Now with the above δ we can find an N such that if $n > N$,

$$| y(x, h_n) - y(x) | < \delta$$

for all x [since the $y(x, h_n)$ converge uniformly to $y(x)$]. Hence, given an $\varepsilon > 0$ there exists an N such that if $n > N$

$$| f(y(x, h_n), x) - f(y(x), x) | < \varepsilon$$

for all $x \varepsilon \mathfrak{R}$.

Since the integral of the limit of a sequence of uniformly convergent continuous functions is equal to the limit of the integral, we have

$$\lim_{n\to\infty}\int_{x_0}^{x} f(y(t, h_n), t)\, dt = \int_{x_0}^{x} f(y(t), t)\, dt. \tag{11}$$

Hence, from Equation (8)

$$\lim_{n\to\infty} y(x, h_n) = \lim_{n\to\infty} y_0 + \lim_{n\to\infty}\int_{x_0}^{x} f(y(t, h_n), t)\, dt + \lim_{n\to\infty} u(x, h_n)$$

and

$$y(x) = y_0 + \int_{x_0}^{x} f(y(t), t)\, dt + 0$$

which is the desired result. [The first limit by the hypothesis of the lemma, the second since y_0 is a constant, the third by Equation (11), and the fourth by Equation (10).]

1.6 Thus we see that in order to obtain a solution of Equation (2) and hence of Equation (1) we must choose a sequence $h_1, h_2, \ldots$ such that $\lim_{n\to\infty} h_n = 0$ and $y(x, h_n)$ converges uniformly to some limit function $y(x)$. In order to do this certain properties of $y(x, h)$ will be established in Lemma 5. Notice now that questions two and three have been transformed into the question of uniform convergence of the polygonal line functions. There are many methods of obtaining approximants and setting up the existence theory; but invariably one passes to an integral equation and requires uniform convergence in order to answer these two questions. How uniform convergence is obtained may vary, but an equivalent discussion to the one used here is usually given.

A convenient concept to use in this connection is that of equicontinuity.

Definition. A sequence of continuous functions $\{f_n(x)\}$ defined on an interval I of the x-axis is said to be *equicontinuous* at $x_0 \varepsilon I$ if, given an $\varepsilon > 0$ there exists a $\delta = \delta(x_0) > 0$ such that if $|x - x_0| < \delta$, then $|f_n(x) - f_n(x_0)| < \varepsilon$ for every n. If δ is independent of x_0, we shall say $\{f_n(x)\}$ is *uniformly equicontinuous* on I.

Lemma 5. (i) The $y(x, h)$ are bounded by a constant C independent of x and h on the interval $\mathfrak{R}$, $|x - x_0| \leq b$.

(ii) The $y(x, h)$ are uniformly equicontinuous.

Proof. Consider Equation (6). This yields

$$|y(x, h)| \leq |y_0| + \left| \int_{x_0}^{x} F(t, h)\, dt \right| \leq |y_0| + M|x - x_0|$$
$$\leq |y_0| + Mb = C \tag{12}$$

and C is independent of x and h.

To prove (ii), write Equation (6) with $x = x'$ and with $x = x''$ and subtract.

$$|y(x', h) - y(x'', h)| = \left| \int_{x''}^{x'} F(t, h)\, dt \right| \leq M|x' - x''|. \tag{13}$$

Now, given an $\varepsilon > 0$, let $\delta = \varepsilon/M$ (independent of x). Clearly, Equation (13) implies (ii).

1.7 We can now proceed to construct a sequence $\{h_n\}$ such that $\{y(x, h_n)\}$ approaches some y(x) uniformly. The basis for this construction depends on a number of lemmas which can advantageously be formulated in somewhat more general terms than necessary for our immediate purposes.

Definition. A sequence of functions $\{f_n(x)\}$ defined on an interval I of the x-axis is said to be *bounded* if there exists a constant M independent of x and n such that $|f_n(x)| \leq M$ for all $x \varepsilon I$ and all n.

Lemma 6. Let $\{f_n(x)\}$ be a bounded sequence of functions defined on a closed finite interval I of the x-axis. Let $x_1, x_2, \ldots$ be a denumerable sequence of points in I. Then there exists a subsequence of $\{f_n(x)\}$ which converges for every x_i.

Proof. Consider the first point x_1 and the sequence

$$f_1(x_1), f_2(x_1), f_3(x_1), \ldots .$$

By hypothesis $|f_n(x)| \leq M$. Since the set of points $[-M, M]$ is compact, there exists a subsequence

$$f_{11}(x), f_{12}(x), f_{13}(x), \ldots$$

of $\{f_n(x)\}$ which converges at x_1. Consider now

$$f_{11}(x_2), f_{12}(x_2), f_{13}(x_2), \ldots .$$

Since this is also a bounded sequence of numbers, $|f_{1j}(x_2)| \leq M$, there exists a subsequence of $\{f_{1j}(x)\}$

$$f_{21}(x), f_{22}(x), f_{23}(x), \ldots$$

which converges at x_2. We continue this process for $x_3, x_4, x_5, \ldots .$

Consider now the diagonal sequence

$$f_{11}(x), f_{22}(x), f_{33}(x), \ldots .$$

Since it is a subset of $\{f_{1j}\}$ it converges at x_1. The sequence also converges at x_1 and x_2 since

$$f_{22}(x), f_{33}(x), f_{44}(x), \ldots$$

is a subsequence of both $\{f_{2j}\}$ and $\{f_{1j}\}$. Continuing in this manner we see (since we can discard any finite number of terms), that

$$f_{11}(x), f_{22}(x), f_{33}(x), \ldots$$

converges for every x_k.

Lemma 7. If $\{f_n(x)\}$ is equicontinuous on a closed finite interval I of the x-axis and convergent on a dense subset I_1 in I, then $\{f_n(x)\}$ converges at every point of I.

Proof. Let x_0 be an arbitrary point in I. We shall show that $\{f_n(x_0)\}$ is a convergent sequence. Given an $\varepsilon > 0$ we can find a $\delta > 0$ such that if $|x - x_0| < \delta$, then for every n

$$|f_n(x) - f_n(x_0)| < \frac{\varepsilon}{3}$$

by equicontinuity of the $\{f_n(x)\}$ on I.

Since I_1 is dense in I there exists an $x_1 \varepsilon I_1$ such that $|x_1 - x_0| < \delta$. Now $\{f_n(x_1)\}$ is convergent by hypothesis. Hence there exists an N such that if m, n > N,

$$|f_n(x_1) - f_m(x_1)| < \frac{\varepsilon}{3}.$$

Now

$$f_n(x_0) - f_m(x_0) = f_n(x_0) - f_n(x_1) + f_n(x_1) - f_m(x_1) + f_m(x_1) - f_m(x_0)$$

and if m, n > N,

$$\begin{aligned} |f_n(x_0) - f_m(x_0)| &\leq |f_n(x_0) - f_n(x_1)| + |f_n(x_1) - f_m(x_1)| \\ &\quad + |f_m(x_1) - f_m(x_0)| \\ &< \frac{\varepsilon}{3} + \frac{\varepsilon}{3} + \frac{\varepsilon}{3} = \varepsilon. \end{aligned}$$

(The first and third by equicontinuity and the second by our choice of N.)

Lemma 8. If $\{f_n(x)\}$ is equicontinuous and convergent on a closed bounded set I, then $\{f_n(x)\}$ converges uniformly on I.

Proof. Let $\varepsilon > 0$ be assigned. For each $x \varepsilon I$ there exists a δ—neighborhood of x such that if $| x' - x | < \delta$, then $| f_n(x') - f_n(x) | < \varepsilon/3$ for every n (by equicontinuity). Since I is compact, there exists, by the Heine-Borel Theorem, a finite number of such δ-neighborhoods (say those associated with $x_1, x_2, \ldots, x_r$) which cover the whole interval I. For every such x_j, $j = 1, 2, \ldots, r$, there is an integer N_j such that if $n, m \geq N_j$,

$$| f_n(x_j) - f_m(x_j) | < \frac{\varepsilon}{3}$$

(since the sequence $\{f_n(x)\}$ is convergent). Let

$$N = \max_j N_j.$$

Now if x is any point of I, it must be in the δ-neighborhood of some x_j, $j = 1, 2, \ldots, r$. Hence if $n, m \geq N \geq N_j$,

$$\begin{aligned} | f_n(x) - f_m(x) | &= | f_n(x) - f_n(x_j) + f_n(x_j) - f_m(x_j) + f_m(x_j) - f_m(x) | \\ &\leq | f_n(x) - f_n(x_j) | + | f_n(x_j) - f_m(x_j) | + | f_m(x_j) - f_m(x) | \\ &< \frac{\varepsilon}{3} + \frac{\varepsilon}{3} + \frac{\varepsilon}{3} = \varepsilon. \end{aligned}$$

(The first and third by equicontinuity, the second by our choice of N.)

1.8 We are now in a position to prove Theorem 1 which was stated in Section 1.1. One begins with the sequence of functions $\{y(x, h_n)\}$ with $h_n = b/n$. Clearly $\lim_{n \to \infty} h_n = 0$. By Lemma 5, $\{y(x, h_n)\}$ is a bounded sequence of equicontinuous functions. By Lemma 6 $\{y(x, h_n)\}$ has a subsequence which converges on the set of rational points in the interval $| x - x_0 | \leq b$. (The set of rational points is denumerable.) By Lemma 7, this subsequence converges throughout the interval $| x - x_0 | \leq b$. (The rationals are dense in the reals.) By Lemma 8, this subsequence converges uniformly, and, since the $y(x, h_n)$ are continuous, the limit function $y(x)$ is continuous. By Lemma 4, this limit function $y(x)$ satisfies Equation (2). By Lemma 2, $y(x)$ is the desired solution of Equation (1).

The continuity of $f(y, x)$ has been used in the above for three purposes. It has been used to show: (1) That $f(y, x)$ is bounded, (2) In Lemma 3 that $u(x, h) \to 0$ as $h \to 0$, and (3) That $f(y(x, h_n), x)$ approaches $f(y(x), x)$ uniformly.

It is also important to notice the significance of the constant b which appears in the conclusion of Theorem 1. This quantity is obtained by a local construction which involves no long range property of f(y, x). In this sense, Theorem 1 is a theorem "in the small." One might suspect that if f(y, x) is continuous everywhere, this b is superfluous, that is, that there is a better theorem which proclaims that there is a solution defined for all values of x. Unfortunately this is not true. For suppose $f(y, x) = y^2$, that is, our differential equation is

$$\frac{dy}{dx} = y^2.$$

Clearly $y = 0$ is a solution defined for every x, but if y is not zero we have

$$\frac{dy/dx}{y^2} = 1$$

and

$$-\frac{1}{y} = x - c$$

or

$$y = \frac{1}{c - x}.$$

Thus if $x_0 = 0, y_0 = 1$ are the initial conditions, $y = \frac{1}{1-x}$ which has only a finite range in which it is continuous — even though our differential equation is continuous everywhere in x and y.

Later we shall see that if f(y, x) is linear in y, existence theorems in the large may be given.

2. The Peano theorem for n functions

2.1 We shall now prove the Peano existence theorem for a system of n equations.

Theorem 2. Let $f_i(y_1, \ldots, y_n, x)$, $i = 1, 2, \ldots, n$ be n real valued functions of the $n + 1$ real variables $y_1, \ldots, y_n, x$ defined and continuous on an open region $\mathfrak{A}$ of $(n + 1)$-dimensional euclidean space. Then for every point $(y_{1,0}, \ldots, y_{n,0}, x_0)$ of $\mathfrak{A}$ we can find a $b > 0$ and n functions $\varphi_1(x), \ldots, \varphi_n(x)$ which have continuous first derivatives in the neighborhood $\mathfrak{N}$, $|x - x_0| \leq b$ such that

$$\frac{d\varphi_i}{dx} = f_i(\varphi_1(x), \ldots, \varphi_n(x), x) \quad i = 1, 2, \ldots, n, \tag{14}$$

in this neighborhood $\mathfrak{N}$ and $\varphi_i(x_0) = y_{i,0}, \quad i = 1, 2, \ldots, n.$

2.2 The proof of this theorem closely parallels the proof given in Section 1 for the case $n = 1$. Our first task is to show the equivalence of the system of differential equations of Equation (14) with the integral form

$$y_i(x) = y_{i,0} + \int_{x_0}^{x} f_i(y_1(t), \ldots, y_n(t), t)\, dt, \quad i = 1, 2, \ldots, n, \tag{15}$$

for $x \varepsilon \mathfrak{N}$.

Suppose for the moment that there exist n functions $y_1(x), \ldots, y_n(x)$ defined for $x \varepsilon \mathfrak{N}$ which satisfy Equation (14) and at $x = x_0$ assume the values $y_{1,0}, \ldots, y_{n,0}$ respectively. Of necessity dy_i/dx must exist and hence $y_i(x)$ is continuous. Since f_i is continuous by hypothesis, $f_i(y_1, \ldots, y_n, x)$ is a continuous function of x in the neighborhood $\mathfrak{N}$ of $x = x_0$. Hence, since $\frac{dy_i}{dx} = f_i(y_1, \ldots, y_n, x)$, $\frac{dy_i}{dx}$ is continuous and consequently,

$$y_i(x) - y_i(x_0) = \int_{x_0}^{x} \frac{dy_i}{dt}\, dt.$$

This equation together with Equation (14) implies Equation (15).

Conversely, n functions $y_i(x)$ satisfying Equation (15) must be continuous functions of x. For, if the $y_i(x)$ satisfy Equation (15), it implies that $f_i(y_1, \ldots, y_n, x)$ is Riemann integrable on $\mathfrak{N}$ and hence $y_{i,0} + \int_{x_0}^{x} f_i(y_1(t), \ldots, y_n(t), t)\, dt$ is continuous. But this expression is precisely $y_i(x)$. Now, since $y_i(x)$ is continuous and the f_i are continuous by hypothesis, $f_i(y_1(t), \ldots, y_n(t), t)$ is continuous for $t \varepsilon \mathfrak{N}$. Thus the derivative of the integral on the right hand side of Equation (15) exists and equals the integrand. Consequently, differentiating Equation (15) will yield Equation (14).

We have thus proved the following lemmas.

Lemma 9. If there exist n functions $\varphi_i(x)$ defined for $x \varepsilon \mathfrak{N}$ which satisfy Equation (14), then $d\varphi_i/dx$ exists and is continuous.

Lemma 10. The solutions $\varphi_i(x)$ of Equation (14) which at $x = x_0$ equal $y_{i,0}$ are identical with the solutions of Equation (15).

The study of the differential equations of Equation (14) has thus been reduced to the study of the integral equations of Equation (15).

2.3 It is necessary to construct the neighborhood $\mathfrak{N}$ of x_0, namely $|x - x_0| \leq b$ which appears in the conclusion of Theorem 2.

Since $\mathfrak{A}$ is an open region, there exists a neighborhood $\mathfrak{R}$ of the point $(y_{1,0}, \ldots, y_{n,0}, x_0)$, namely $|y_i - y_{i,0}| \leq a$, $|x - x_0| \leq a$ included in the region $\mathfrak{A}$. Since $\mathfrak{R}$ is a closed point set interior to $\mathfrak{A}$ and the $f_i(y_1, \ldots, y_n, x)$ are continuous on $\mathfrak{A}$, the f_i are bounded on $\mathfrak{R}$. One can, of course, find a common bound M for the n functions f_i such that for $(y_1, \ldots, y_n, x) \varepsilon \mathfrak{R}$

$$|f_i(y_1, \ldots, y_n, x)| \leq M, \quad i = 1, 2, \ldots, n.$$

Now as before let

$$b = \min\left(a, \frac{a}{M}\right)$$

and consider the set of points

$$|y_i - y_{i,0}| \leq a, \qquad |x - x_0| \leq b.$$

We shall call this closed $(n + 1)$-dimensional rectangular region of $\mathfrak{R}$, $\mathfrak{R}'$. Clearly,

(i) $f_i(y_1, \ldots, y_n, x)$ is continuous in $\mathfrak{R}'$ with bound M.

(ii) If $|x - x_0| \leq b$ and $|y_i - y_{i,0}| \leq M|x - x_0|$, then $(y_1, \ldots, y_n, x) \varepsilon \mathfrak{R}'$.

To prove (i) we merely note that $\mathfrak{R}' \leq \mathfrak{R} < \mathfrak{A}$. To demonstrate (ii), we note that

$$|y_i - y_{i,0}| \leq M|x - x_0| \leq Mb \leq M\frac{a}{M} = a.$$

We now proceed to construct n polygonal line functions which are intended to approximate the n solutions of Equation (14). Let $h > 0$ be assigned. It is convenient to suppose that b is an integral multiple of h, that is,

$$b = qh$$

for some positive integer q.

As before we can define q additional points $\{y_{1j}, \ldots, y_{nj}, x_j\}$, $j = 1, 2, \ldots, q$ inductively by the equations

$$y_{ij} = y_{i,j-1} + f_i(y_{1,j-1}, \ldots, y_{n,j-1}, x_{j-1})\,h, \quad j = 1, 2, \ldots, q$$
$$x_j = x_{j-1} + h$$

showing inductively that $f_i(y_{1j}, \ldots, y_{nj}, x_j)$ is in $\mathfrak{R}'$ by means of (i) and (ii) above.

Also, define the functions $y_i(x, h)$ by the equations

$$y_i(x, h) = y_{i,j-1} + f_i(y_{1,j-1}, \ldots, y_{n,j-1}, x_{j-1})(x - x_{j-1}), \qquad i = 1, 2, \ldots, n \tag{16}$$

for

$$0 < x - x_{j-1} \leq h$$

and

$$j = 1, 2, \ldots, q$$

The quantity $y_i(x_0, h)$ is defined as

$$y_i(x_0, h) = y_{i,0}.$$

Note that x_0 and the $y_{i,0}$ are the initial point of Theorem 2.

The functions $y_i(x, h)$ have thus been defined for all x in the interval $x_0 \leq x \leq x_0 + b$. They are continuous functions of x, each consisting of a finite number of straight line segments. A similar definition of $y_i(x, h)$ can be given for the interval $x_0 - b \leq x < x_0$ by letting $y_i(x, h)$ equal $y_{i,j+1} + f_i(y_{1,j+1}, \ldots, y_{n,j+1}, x_{j+1})(x - x_{j+1})$ for $-h \leq x - x_{j+1} < 0$ where $y_{ij} = y_i(x_j, h)$, $x_j = x_0 + jh$ and $j = -1, -2, \ldots, -q$.

For simplicity in notation set f_{ik} equal to

$$f_i(y_{1k}, \ldots, y_{nk}, x_k), \qquad i = 1, 2, \ldots, n.$$

Now in the region $x_0 \leq x \leq x_0 + b$ define the functions $F_i(x, h)$ by the equations

$$F_i(x, h) = f_{i,j-1}, \qquad i = 1, 2, \ldots, n \tag{17}$$

for $0 < x - x_{j-1} \leq h$, $j = 1, 2, \ldots, q$ and in the region $x_0 - b \leq x < x_0$ let the $F_i(x, h)$ be defined by

$$F_i(x, h) = f_{i,j+1}, \qquad i = 1, 2, \ldots, n \tag{18}$$

for $-h \leq x - x_{j+1} < 0$, $j = -1, -2, \ldots, -q$.

From Equation (16) we conclude that

$$y_i(x, h) = y_{i,0} + \sum_{k=0}^{j-2} f_{ik}h + f_{i,j-1}(x - x_{j-1}) \tag{19}$$

$$j > 1, \quad 0 < x - x_{j-1} \leq h.$$

Also

$$y_i(x, h) = y_{i,0} - \sum_{k=0}^{j+2} f_{ik}\, h + f_{i,j+1}(x - x_{j+1}) \tag{20}$$

$$j < -1, \quad -h \leq x - x_{j+1} < 0$$

while

$$y_i(x, h) = y_{i,0} + f_{i0}(x - x_0), \quad -h \leq x - x_0 \leq h. \tag{21}$$

From Equations (17), (18), (19), (20) and (21) we see that

$$y_i(x, h) = y_{i,0} + \int_{x_0}^{x} F_i(t, h)\, dt, \quad i = 1, 2, \ldots, n \tag{22}$$

for $x \in \mathfrak{R}$, that is, $x_0 - b \leq x \leq x_0 + b$.

We now recall, as in the case $n = 1$, that the points $(y_{1j}, \ldots, y_{nj}, x_j)$ are in $\mathfrak{R}'$ (and hence the arguments of the F_i are in our region of definition); and that $| F_i(t, h) | \leq M$ for $| x - x_0 | \leq b$. But from this result and Equation (22)

$$| y_i(x, h) - y_{i,0} | \leq M \, | x - x_0 |$$

which by virtue of (ii) above implies

$$(y_1(x, h), \ldots, y_n(x, h), x) \in \mathfrak{R}'$$

for all $x \in \mathfrak{R}$.

Since the $F_i(x, h)$ are step functions with a finite number of jumps, Equation (22) defines continuous functions of x and we may write

$$y_i(x, h) = y_{i,0} + \int_{x_0}^{x} f_i(y_1(t, h), \ldots, y_n(t, h), t)\, dt + u_i(x, h), \quad i = 1, 2, \ldots, n \tag{23}$$

for $x \in \mathfrak{R}$ where, by definition,

$$u_i(x, h) = \int_{x_0}^{x} [F_i(t, h) - f_i(y_1(t, h), \ldots, y_n(t, h), t)]\, dt. \tag{24}$$

2.4 To further our attempt to prove that the polygonal functions $y_i(x, h)$ are approximations to the solutions of Equation (14) we prove Lemma 11 below. First we recall that since $\mathfrak{R}$ is compact and $f_i(y_1, \ldots, y_n, t)$ is continuous on $\mathfrak{R}$, f_i is uniformly continuous. Thus, given an $\varepsilon > 0$ there exists a $\delta_i > 0$ such that

$$| f_i(y_1', \ldots, y_n', x') - f_i(y_1'', \ldots, y_n'', x'') | < \varepsilon$$

for all $(y_1', \ldots, y_n', x')$ and $(y_1'', \ldots, y_n'', x'')$ in $\mathfrak{R}$ with $| x' - x'' | < \delta_i$, $| y_j' - y_j'' | < \delta_i$. Thus there exists a uniform δ_i for each f_i. Now for a given ε one can take $\delta = \min_i \delta_i$ as the "uniform δ for the f_i."

Lemma 11. Let $\varepsilon > 0$ be assigned. Let $\delta > 0$ be the uniform δ for $\frac{\varepsilon}{b}$ and let $h_0 = \min\left(\delta, \frac{\delta}{M}\right)$. Then if $h < h_0$ and $|x - x_0| \leq b$,

$$|u_i(x, h)| \leq \varepsilon, \quad i = 1, 2, \ldots, n.$$

Proof. Suppose $x \geq x_0$ and $0 < x - x_{j-1} \leq h$. From Equation (16)

$$\begin{aligned} |y_i(x, h) - y_{i,j-1}| &= |f_i(y_{1,j-1}, \ldots, y_{n,j-1}, x_{j-1})|\,|x - x_{j-1}| \\ &\leq M\,|x - x_{j-1}| \leq Mh < Mh_0 \leq \delta. \end{aligned}$$

Also

$$\begin{aligned} \frac{\varepsilon}{b} &> |f_i(y_1(x, h), \ldots, y_n(x, h), x) - f_i(y_{1,j-1}, \ldots, y_{n,j-1}, x_{j-1})| \\ &= |f_i(y_1(x, h), \ldots, y_n(x, h), x) - F_i(x, h)| \end{aligned}$$

and

$$\begin{aligned} |u_i(x, h)| &\leq \left| \int_{x_0}^{x} |F_i(t, h) - f_i(y_1(t, h), \ldots, y_n(t, h), t)|\, dt \right| \\ &< \frac{\varepsilon}{b}|x - x_0| \leq \varepsilon. \end{aligned} \tag{25}$$

A similar argument holds for $x_0 - b \leq x < x_0$.

2.5 As $h \to 0$ it appears that our polygonal functions will approach solutions of Equation (14) as in the case of $n = 1$. We continue the development analogous to that of the earlier case by proving Lemma 12.

Lemma 12. Let $h_1, h_2, \ldots$ be a sequence of positive numbers such that $\lim_{r \to \infty} h_r = 0$. Suppose that for $|x - x_0| \leq b$, the $y_i(x, h_r)$ converge uniformly to functions $y_i(x)$ for each $i = 1, 2, \ldots, n$. Then for $|x - x_0| \leq b$

$$y_i(x) = y_{i,0} + \int_{x_0}^{x} f_i(y_1(t), \ldots, y_n(t), t)\, dt, \quad i = 1, 2, \ldots, n.$$

Proof. Consider Equation (23) with h replaced by h_r. By Lemma 11,

$$u_i(x, h_r) \to 0, \quad i = 1, 2, \ldots, n$$

uniformly for any $x \in \mathfrak{R}$. Also, since the $y_i(x, h_r)$ converge uniformly to the $y_i(x)$ by hypothesis, the $y_i(x)$ are continuous, and since the f_i are uniformly continuous

$$\lim_{r \to \infty} f_i(y_1(x, h_r), \ldots, y_n(x, h_r), x) = f_i(y_1(x), \ldots, y_n(x), x)$$

for $i = 1, 2, \ldots, n$ uniformly in x. Since the integral of the limit of a sequence of uniformly converging functions is equal to the limit of the integral, we have

$$\lim_{r\to\infty}\int_{x_0}^{x} f_i(y_1(t,h_r),\ldots,y_n(t,h_r),t)\,dt = \int_{x_0}^{x} f_i(y_1(t),\ldots,y_n(t),t)\,dt. \quad (26)$$

Hence from Equation (23) we conclude that

$$y_i(x) = y_{i,0} + \int_{x_0}^{x} f_i(y_1(t), \ldots, y_n(t), t)\,dt.$$

2.6 As before, we see that a sequence of h's must be chosen such that $\lim_{r\to\infty} h_r = 0$ and $y_i(x, h_r)$ converge uniformly.

Lemma 13. (i) The $y_i(x, h)$ are bounded by a constant C independent of x, h and i on the interval $\mathfrak{R}$, $|x - x_0| \leq b$.

(ii) The $y_i(x, h)$ are uniformly equicontinuous.

Proof. From Equation (22)

$$|y_i(x, h)| \leq |y_{i,0}| + \left|\int_{x_0}^{x} F_i(t, h)\,dt\right| \leq |y_{i,0}| + M\,|x - x_0|$$
$$\leq |y_{i,0}| + Mb = C_i.$$

Let $C = \max_i C_i$. Clearly, C is independent of x, h and i.

To prove (ii), write Equation (22) with $x = x'$ and with $x = x''$ and subtract.

$$|y_i(x', h) - y_i(x'', h)| = \left|\int_{x''}^{x'} F_i(t, h)\,dt\right| \leq M\,|x' - x''|. \quad (27)$$

Now, given an $\varepsilon > 0$, let $\delta = \varepsilon/M$ (independent of x). Clearly Equation (27) implies (ii).

2.7 We are now in a position to prove Theorem 2 as stated in Section 2.1. (Lemmas 6, 7, and 8 of Section 1 will also be used.)

One begins with the n sequences of functions $\{y_i(x, h_r)\}$ with $h_r = \frac{b}{r}$. Clearly $\lim_{r\to\infty} h_r = 0$. By Lemma 13 (i) the $\{y_i(x, h_r)\}$ are bounded sequences of functions. By Lemma 6 there exists a subsequence of $\{y_1(x, h_r)\}$, say $\{y_1(x, h_{rs_1})\}$ which converges on a dense set of points in $|x - x_0| \leq b$. By Lemma 13 (ii), Lemma 7 and

Lemma 8, $\{y_1(x, h_{rs_1})\}$ converges uniformly to a continuous function $y_1(x)$, $x \varepsilon \mathfrak{R}$. One next considers the sequence $\{y_2(x, h_{rs_1})\}$ and, as indicated above, we may extract a subsequence $\{y_2(x, h_{rs_2})\}$ which converges uniformly on $\mathfrak{R}$. Continuing this process of selecting subsequences one eventually obtains a sequence of h's, namely $\{h_{rs_n}\}$ such that $\{y_i(x, h_{rs_n})\}$ converges uniformly to a continuous function $y_i(x)$ for all $i = 1, 2, \ldots, n$. By Lemma 12, the $y_i(x)$ satisfy Equation (15) and by Lemma 10, the $y_i(x)$ are the desired solutions of Equation (14).

CHAPTER 2

General Existence Theorems

1. The implicit function theorem

1.1 In the previous chapter we stated and proved an existence theorem for a system of n differential equations

$$\frac{dy_i}{dx} = f_i(y_1, \ldots, y_n, x), \qquad i = 1, 2, \ldots, n. \tag{1}$$

Frequently one has to deal with systems of differential equations not explicitly solved for the derivatives, say

$$F_i\left(\frac{dy_1}{dx}, \ldots, \frac{dy_n}{dx}, y_1, \ldots, y_n, x\right) = 0, \quad i = 1, \ldots, n. \tag{2}$$

By means of the implicit function theorem, suitable systems of the form of Equation (2) can be reduced to Equation (1) for which the Peano existence theorem applies. The implicit function theorem will also permit us to deduce various other properties of the functions f_i which will be useful in the future. For these reasons we shall discuss the implicit function theorem here.

1.2 The implicit function theorem does not, of course, involve the idea of a derivative; however, if one likes, one may consider it a theorem for differential equations of order zero. In the present section we shall state the theorem in its general form, while later sections will be devoted to its proof.

A number of interrelated results will be proved in the form of lemmas. Common to all these lemmas we have a basic hypothesis.

Hypothesis. H 1. Let

$$F_i(z_1, \ldots, z_n, x_1, \ldots, x_m), \quad i = 1, \ldots, n \tag{3}$$

be a set of n real valued functions of (n + m) real variables defined and continuous on an open region $\mathfrak{A}$ of (n + m)-dimensional euclidean space.

H 2. The F_i have continuous first derivatives relative to the z_j, $j = 1, \ldots, n$ at every point $(z_1, \ldots, z_n, x_1, \ldots, x_m)$ of $\mathfrak{A}$.

H 3. Let $(z_{1,0}, \ldots, z_{n,0}, x_{1,0}, \ldots, x_{m,0})$ be a point of $\mathfrak{A}$ such that

$$F_i(z_{1,0}, \ldots, z_{n,0}, x_{1,0}, \ldots, x_{m,0}) = 0, \quad i = 1, \ldots, n \tag{4}$$

and the Jacobian

$$J = \frac{\partial(F_1, \ldots, F_n)}{\partial(z_1, \ldots, z_n)} = \begin{vmatrix} \frac{\partial F_1}{\partial z_1} & \frac{\partial F_1}{\partial z_2} & \cdots & \frac{\partial F_1}{\partial z_n} \\ \frac{\partial F_2}{\partial z_1} & \frac{\partial F_2}{\partial z_2} & \cdots & \frac{\partial F_2}{\partial z_n} \\ \cdot & \cdot & \cdot & \cdot \\ \frac{\partial F_n}{\partial z_1} & \frac{\partial F_n}{\partial z_2} & \cdots & \frac{\partial F_n}{\partial z_n} \end{vmatrix} \tag{5}$$

is not zero at this point.

The implicit function theorem may now be stated.

Theorem 1. Under the basic hypothesis H1, H2, H3 above there exists a positive number b' and n unique continuous functions $\varphi_r(x_1, \ldots, x_m)$ such that $z_{r,0} = \varphi_r(x_{1,0}, \ldots, x_{m,0})$ and

$$F_i(\varphi_1, \ldots, \varphi_n, x_1, \ldots, x_m) \equiv 0, \quad i = 1, \ldots, n \tag{6}$$

for $| x_i - x_{i,0} | \leq b'$.

1.3 Our first result towards proving the implicit function theorem is given by Lemma 1.

Lemma 1. Under the hypothesis H1, H2, H3 there exists a rectangular neighborhood $\mathfrak{B}$ of the $x_{i,0}$ in m-dimensional euclidean space and a rectangular neighborhood $\mathfrak{C}$ of the $z_{j,0}$ in n-dimensional euclidean space such that to each point $(x_1', \ldots, x_m')$ of $\mathfrak{B}$ there exists at most one point $(z_1', \ldots, z_n')$ of $\mathfrak{C}$ such that

$$F_i(z_1', \ldots, z_n', x_1', \ldots, x_m') = 0, \quad i = 1, \ldots, n. \tag{7}$$

Proof. Since $\mathfrak{A}$ is an open region there exists an $a > 0$ such that the rectangular neighborhood

$$| z_j - z_{j,0} | \leq a, \qquad | x_i - x_{i,0} | \leq a$$

lies entirely in $\mathfrak{A}$. If we call the totality of points

$$| x_i - x_{i,0} | \leq a \quad \text{(in m-dimensional space)} \tag{8}$$

the region $\mathfrak{B}^*$ and the totality of points

$$| z_j - z_{j,0} | \leq a \quad \text{(in n-dimensional space)} \tag{9}$$

the region $\mathfrak{C}^*$, the region $\mathfrak{C}^* \times \mathfrak{B}^*$ in $(n + m)$-dimensional space [consisting of all points $(z_1, \ldots, z_n, x_1, \ldots, x_m)$ where the x_i and z_j satisfy Equations (8) and (9)] is clearly in $\mathfrak{A}$.

Consider now the Jacobian of Equation (5). J is, of course, to be regarded as a function of the $n + m$ variables $z_1, \ldots, z_n, x_1, \ldots, x_m$. We now consider a determinant K similar to J except that the variables in different rows are different,

$$K = \begin{vmatrix} \frac{\partial F_1}{\partial z_1}(z_{11}, \ldots, z_{n1}, x_{11}, \ldots, x_{m1}) \ldots \frac{\partial F_1}{\partial z_n}(z_{11}, \ldots, z_{n1}, x_{11}, \ldots, x_{m1}) \\ \frac{\partial F_2}{\partial z_1}(z_{12}, \ldots, z_{n2}, x_{12}, \ldots, x_{m2}) \ldots \frac{\partial F_2}{\partial z_n}(z_{12}, \ldots, z_{n2}, x_{12}, \ldots, x_{m2}) \\ \cdots\cdots\cdots\cdots\cdots\cdots\cdots\cdots \\ \frac{\partial F_n}{\partial z_1}(z_{1n}, \ldots, z_{nn}, x_{1n}, \ldots, x_{mn}) \ldots \frac{\partial F_n}{\partial z_n}(z_{1n}, \ldots, z_{nn}, x_{1n}, \ldots, x_{mn}) \end{vmatrix}.$$

The determinant K is a function of the $n(n + m)$ variables z_{jh}, x_{ih}, $j = 1, \ldots, n$, $h = 1, \ldots, n$, $i = 1, \ldots, m$. When $z_{jh} = z_j$, $x_{ih} = x_i$ for all h, K becomes identical with J. At the point

$$\begin{array}{lll} z_{jh} = z_{j,0}, & j = 1, \ldots, n & \\ & & h = 1, \ldots, n \\ x_{ih} = x_{i,0}, & i = 1, \ldots, m & \end{array}$$

K is unequal to zero by hypothesis H3. Let K_0 be this value of K. For values of z_{jh} and x_{ih} such that

$$| z_{jh} - z_{j,0} | \leq a, \qquad | x_{ih} - x_{i,0} | \leq a$$

each $\partial F_i/\partial z_k$ is a continuous function of its $n + m$ variables and hence K is a continuous function of its $n(n + m)$ variables. Therefore, we can find a region $\mathfrak{B}$ in $n(n + m)$-dimensional space,

$$\begin{array}{l} | z_{jh} - z_{j,0} | \leq b \leq a \\ | x_{ih} - x_{i,0} | \leq b \leq a \end{array}$$

for which

$$|K - K_0| \leq \tfrac{1}{2}|K_0|.$$

(b is the "δ" and $\frac{1}{2}|K_0|$ the "ε"). Hence for this region 𝔅

$$|K| \geq \tfrac{1}{2}|K_0|.[1]$$

We shall now show that if there exists a point $(x_1', \ldots, x_m')$ such that

$$|x_i' - x_{i,0}| \leq b$$

and two points $(z_1', \ldots, z_n')$ and $(z_1'', \ldots, z_n'')$ both of which satisfy the inequalities

$$|z_j' - z_{j,0}| \leq b, \qquad |z_j'' - z_{j,0}| \leq b$$

for which

$$F_i(z_1', \ldots, z_n', x_1', \ldots, x_m') = 0, \qquad i = 1, \ldots, n \tag{10}$$

and

$$F_i(z_1'', \ldots, z_n'', x_1', \ldots, x_m') = 0, \qquad i = 1, \ldots, n \tag{11}$$

then

$$z_j' = z_j'', \qquad j = 1, \ldots, n.$$

Subtract Equation (11) from Equation (10) for corresponding values of i. Because of H2 in the basic hypothesis we may apply the Law of the Mean for n variables,

$$F_i(z_1', \ldots, z_n', x_1', \ldots, x_m') - F_i(z_1'', \ldots, z_n'', x_1', \ldots, x_m') = 0$$

$$= \sum_{r=1}^{n} \frac{\partial F_i}{\partial z_r}(z_{1i}^*, \ldots, z_{ni}^*, x_1', \ldots, x_m')(z_r' - z_r'') \tag{12}$$

where each partial is evaluated at the same point for i fixed. Now z_{ri}^* lies between z_r' and z_r'' and hence

$$|z_{ri}^* - z_{r,0}| \leq b.$$

[1] The "$\frac{1}{2}$" is unimportant. All we desire to prove is that K is bounded away from zero in the neighborhood of the point with $z_{jh} = z_{j,0}$, $x_{jh} = x_{j,0}$. Clearly, if we use the same variables in each row, K becomes identical with J and we have the result that J is also bounded away from zero in the region ℭ × 𝔅 in (n + m)-dimensional space defined by $|z_j - z_{j,0}| \leq b$, $|x_i - x_{i,0}| \leq b$.

Equation (12) can be regarded as n linear homogeneous algebraic equations on $(z'_r - z''_r)$, $r = 1, \ldots, n$ whose determinant K is not zero by construction ($|K| \geq \frac{1}{2}|K_0| > 0$). Hence by Cramer's rule, $z'_r - z''_r = 0$ and the two points $(z'_1, \ldots, z'_n)$ and $(z''_1, \ldots, z''_n)$ are identical.

If we call the totality of points $(x_1, \ldots, x_m)$ for which

$$|x_i - x_{i,0}| \leq b, \qquad i = 1, \ldots, m$$

the region $\mathfrak{B}$ in m-dimensional space and the points $(z_1, \ldots, z_n)$ for which

$$|z_j - z_{j,0}| \leq b, \qquad j = 1, \ldots, n$$

the region $\mathfrak{C}$ in n-dimensional space, the proof of the lemma is complete.

1.4 In Lemma 1 we have shown the existence of *at most* one point $(z_1, \ldots, z_n)$ which satisfies $F_i = 0$ for a given $(x_1, \ldots, x_m)$. In Lemma 2 we shall show the existence of *at least* one point $(z_1, \ldots, z_n)$ with this property.

Lemma 2. There exists a non-empty sub-region $\mathfrak{B}'$ of $\mathfrak{B}$, say the points $(x_1, \ldots, x_m)$ with $|x_i - x_{i,0}| \leq b'$ such that to each point of $\mathfrak{B}'$ there exists a point $(z_1, \ldots, z_n)$ of $\mathfrak{C}$ such that

$$F_i(z_1, \ldots, z_n, x_1, \ldots, x_m) = 0, \qquad i = 1, \ldots, n.$$

Proof. Consider the quadratic form

$$\mu(z_1, \ldots, z_n, x_1, \ldots, x_m) = \sum_{i=1}^{n} F_i^2(z_1, \ldots, z_n, x_1, \ldots, x_m)$$

defined on $\mathfrak{C} \times \mathfrak{B}$. When $x_i = x_{i,0}$, $i = 1, \ldots, m$; μ becomes a function of $z_1, \ldots, z_n$ defined on $\mathfrak{C}$,

$$\mu(z_1, \ldots, z_n, x_{1,0}, \ldots, x_{m,0}) = \sum_{i=1}^{n} F_i^2(z_1, \ldots, z_n, x_{1,0}, \ldots, x_{m,0}).$$

At $z_r = z_{r,0}$, $r = 1, \ldots, n$, $\mu(z_1, \ldots, z_n, x_{1,0}, \ldots, x_{m,0}) = 0$ by H 3 and by Lemma 1 this zero is unique. Hence, in particular, $\mu(z_1, \ldots, z_n, x_{1,0}, \ldots, x_{m,0})$ is not zero on the boundary of $\mathfrak{C}$. Because $\mathfrak{C}$ is a closed point set, $\mu(z_1, \ldots, z_n, x_{1,0}, \ldots, x_{m,0})$ is bounded away from zero on the boundary of $\mathfrak{C}$; that is

$$\mu(z_1, \ldots, z_n, x_{1,0} \ldots, x_{m,0}) \geq \mu_0 > 0 \tag{13}$$

for some μ_0 and all z on the boundary of $\mathfrak{C}$.

Since $\mathfrak{C} \times \mathfrak{B}$ is closed, μ is uniformly continuous on $\mathfrak{C} \times \mathfrak{B}$. Let us take $\mu_0/2$ as our ε. Then there exists a $\delta > 0$ such that if

$$|z_j' - z_j''| < \delta \qquad \text{and} \qquad |x_i' - x_i''| < \delta$$

with $(z_1', \ldots, z_n', x_1', \ldots, x_m')$ and $(z_1'', \ldots, z_n'', x_1'', \ldots, x_m'')$ in $\mathfrak{C} \times \mathfrak{B}$,

$$|\mu(z_1', \ldots, z_n', x_1', \ldots, x_m') - \mu(z_1'', \ldots, z_n'', x_1'', \ldots, x_m'')| < \mu_0/2.$$

Call the b' of the lemma, δ. Then for $|x_i - x_{i,0}| < b'$ and $(z_1, \ldots, z_n) \varepsilon \mathfrak{C}$ one has

$$|\mu(z_1, \ldots, z_n, x_1, \ldots, x_m) - \mu(z_1, \ldots, z_n, x_{1,0}, \ldots, x_{m,0})| < \mu_0/2. \quad (14)$$

Equation (13) now shows that for $(z_1, \ldots, z_n)$ on the boundary of $\mathfrak{C}$ one has

$$|\mu(z_1, \ldots, z_n, x_1, \ldots, x_m)| > \mu_0/2$$

while for $(z_1, \ldots, z_n) = (z_{1,0}, \ldots, z_{n,0})$, Equation (14) shows

$$|\mu(z_{1,0}, \ldots, z_{n,0}, x_1, \ldots, x_m)| < \mu_0/2.$$

Now choose a point $(x_1', \ldots, x_m')$ in $\mathfrak{B}'$ and hold it fixed. Consider μ for this x point and for any $(z_1, \ldots, z_n) \varepsilon \mathfrak{C}$. On the boundary of $\mathfrak{C}$, $\mu > \mu_0/2$ and at the center point of $\mathfrak{C}$, $\mu < \mu_0/2$. Now μ is a continuous function on a closed point set and hence must have a minimum on $\mathfrak{C}$. By the above inequalities this minimum must be attained at an interior point of $\mathfrak{C}$, say $(z_1', \ldots, z_n')$. Since $\partial\mu/\partial z_r$ exists by H2, $\partial\mu/\partial z_r = 0$ at $(z_1', \ldots, z_n')$ for $r = 1, \ldots, n$. This can be written

$$2\sum_{i=1}^{n} F_i \frac{\partial F_i}{\partial z_r} = 0, \qquad r = 1, \ldots, n \quad (15)$$

where F_i and $\partial F_i/\partial z_r$ are evaluated at $(z_1', \ldots, z_n', x_1', \ldots, x_m')$.

Equation (15) can now be considered as n linear homogeneous algebraic equations on the F_i whose determinant is the Jacobian $J(z_1', \ldots, z_n', x_1', \ldots, x_m')$. Since the K of Lemma 1 is not zero $J(z_1', \ldots, z_n', x_1', \ldots, x_m')$ is not zero. Consequently, by Cramer's rule,

$$F_i(z_1', \ldots, z_n', x_1', \ldots, x_m') = 0. \quad (16)$$

1.5 Combining Lemmas 1 and 2 one has for every point $(x_1, \ldots, x_m)$ with $|x_i - x_{i,0}| \leq b'$ a unique point $(z_1, \ldots, z_n)$ of $\mathfrak{C}$ such that

$$F_i(z_1, \ldots, z_n, x_1, \ldots, x_m) = 0, \qquad i = 1, \ldots, n.$$

Thus for $|x_i - x_{i,0}| \leq b'$ one has functions

$$z_j = \varphi_j(x_1, \ldots, x_m), \qquad j = 1, \ldots, n$$

defined for $(x_1, \ldots, x_m)$ in $\mathfrak{B}'$ such that

$$F_i(\varphi_1, \ldots, \varphi_n, x_1, \ldots, x_m) \equiv 0, \qquad i = 1, \ldots, n. \tag{17}$$

It remains but to show that the φ_j are continuous.

Lemma 3. The functions $\varphi_j(x_1, \ldots, x_m)$ defined above are continuous for $|x_i - x_{i,0}| \leq b'$.

Proof. Let $(x_1', \ldots, x_m')$ and $(x_1'', \ldots, x_m'')$ be two points of $\mathfrak{B}'$, $(z_1', \ldots, z_n')$ and $(z_1'', \ldots, z_n'')$ the corresponding points of $\mathfrak{C}$ such that

$$\left.\begin{array}{l} F_i(z_1', \ldots, z_n', x_1', \ldots, x_m') = 0 \\ F_i(z_1'', \ldots, z_n'', x_1'', \ldots, x_m'') = 0 \end{array}\right. \quad i = 1, \ldots, n.$$

Subtracting the first of these equations from the second and adding and subtracting $F_i(z_1', \ldots, z_n', x_1'', \ldots, x_m'')$ there results

$$\begin{aligned} F_i(z_1'', \ldots, z_n'', x_1'', \ldots, x_m'') - F_i(z_1', \ldots, z_n', x_1'', \ldots, x_m'') \\ + F_i(z_1', \ldots, z_n', x_1'', \ldots, x_m'') - F_i(z_1', \ldots, z_n', x_1', \ldots, x_m') = 0. \end{aligned} \tag{18}$$

Now let

$$F_i(z_1', \ldots, z_n', x_1'', \ldots, x_m'') - F_i(z_1', \ldots, z_n', x_1', \ldots, x_m') = \Delta F_i.$$

Since the F_i have derivatives with respect to the z's by the basic hypothesis, we may apply the Law of the Mean to Equation (18) to obtain,

$$\sum_{r=1}^{n} \frac{\partial F_i}{\partial z_r}(z_{1i}^*, \ldots, z_{ni}^*, x_1'', \ldots, x_m'')(z_r'' - z_r') + \Delta F_i = 0 \tag{19}$$

where z_{ri}^* lies between z_r'' and z_r' but may be different for different i.

If we apply Cramer's rule to Equation (19) and, as before, call K the determinant of the system,

$$z_r'' - z_r' = -\frac{K_r}{K} \tag{20}$$

where K_r is the determinant obtained from K by replacing the r*th* column of K by the column of ΔF_i's. For the range of values of the variables considered, Lemma 1 shows that K is bounded away from zero.

Now hold $(x_1', \ldots, x_m')$ fixed and let $(x_1'', \ldots, x_m'')$ approach $(x_1', \ldots, x_m')$. Then for each i, $\Delta F_i \to 0$ since ΔF_i is a continuous function. Consequently, $K_r \to 0$. Hence

$$\lim_{(x_1'', \ldots, x_m'') \to (x_1', \ldots, x_m')} [\varphi_r(x_1'', \ldots, x_m'') - \varphi_r(x_1', \ldots, x_m')]$$
$$= \lim (z_r'' - z_r') = \lim (-K_r/K) = 0.$$

This is, of course, precisely the statement that φ_r is continuous at $(x_1', \ldots, x_m')$.

Corollary.
$$\lim_{(x_1'', \ldots, x_m'') \to (x_1', \ldots, x_m')} K$$
$$= J(\varphi_1(x_1', \ldots, x_m'), \ldots, \varphi_n(x_1', \ldots, x_m'), x_1', \ldots, x_m').$$

Proof. K depends on values of z_r which lie between z_r'' and z_r' and on the values of x_j'' appearing in the continuous partials $\partial F_i/\partial z_k$. Since $z_r'' \to z_r'$ as $(x_1'', \ldots, x_m'') \to (x_1', \ldots, x_m')$, and the φ_r are continuous, K has as its limit the Jacobian of the F_i evaluated at $(x_1, \ldots, x_m) = (x_1', \ldots, x_m')$.

1.6 The proof of the implicit function theorem is now complete. From Lemmas 1 and 2 there exists one and only one point $(z_1, \ldots, z_n) \varepsilon \mathfrak{C}$ such that for a given $(x_1, \ldots, x_m) \varepsilon \mathfrak{B}'$

$$F_i(z_1, \ldots, z_n, x_1, \ldots, x_m) = 0, \qquad i = 1, \ldots, n.$$

These single valued functions $z_i = \varphi_i(x_1, \ldots, x_m)$ are continuous by Lemma 3.

2. The Peano theorem in the implicit case

The implicit function theorem can be immediately applied to systems of differential equations not explicitly solved for the derivatives.

Theorem 2. Hypothesis. (i) Given the n functions

$$F_i(y_1', \ldots, y_n', y_1, \ldots, y_n, x) = 0,^1 \quad i = 1, \ldots, n$$

[1] We now use the conventional notation

$$y_i' = \frac{dy_i}{dx}.$$

of the $2n + 1$ real variables $y'_1, \ldots, y'_n, y_1, \ldots, y_n, x$ defined and continuous on an open region $\mathfrak{A}$ of $(2n + 1)$-dimensional euclidean space.

(ii) $\partial F_i/\partial y'_j$ exist and are continuous on $\mathfrak{A}$ for $i, j = 1, \ldots, n$.

(iii) There exists a set of values $(y'_{1,0}, \ldots, y'_{n,0}, y_{1,0}, \ldots, y_{n,0}, x_0)$ in $\mathfrak{A}$ such that

$$F_i(y'_{1,0}, \ldots, y'_{n,0}, y_{1,0}, \ldots, y_{n,0}, x_0) = 0, \quad i = 1, \ldots, n \tag{21}$$

and

$$J = \frac{\partial(F_1, \ldots, F_n)}{\partial(y'_1, \ldots, y'_n)} \neq 0$$

at this point.

Conclusion. There exists a positive number b' such that for every set of $n + 1$ values $y^*_1, \ldots, y^*_n, x^*$ with $|y^*_j - y_{j,0}| < b'$ and $|x^* - x_0| < b'$ there exists a $b^* > 0$ $(b^* \leq b')$ and n functions $\varphi^*_1(x), \ldots, \varphi^*_n(x)$ with

$$y^*_j = \varphi^*_j(x^*), \qquad j = 1, \ldots, n \tag{22}$$

which have continuous first derivatives in the neighborhood $|x - x^*| \leq b^*$ such that

$$F_i\left(\frac{d\varphi^*_1}{dx}, \ldots, \frac{d\varphi^*_n}{dx}, \varphi^*_1, \ldots, \varphi^*_n, x\right) \equiv 0, \qquad i = 1, \ldots, n. \tag{23}$$

Proof. The hypothesis of this theorem is precisely the basic hypothesis H1, H2, H3 of the implicit function theorem. Hence there exists a $b' > 0$ such that for $|y_j - y_{j,0}| \leq b'$, $|x - x_0| \leq b'$ there exist n continuous functions

$$y'_j = f_j(y_1, \ldots, y_n, x), \qquad j = 1, \ldots, n \tag{24}$$

such that

$$F_i(f_1, \ldots, f_n, y_1, \ldots, y_n, x) \equiv 0, \quad i = 1, \ldots, n. \tag{25}$$

The constant b' determines a region analogous to $\mathfrak{A}$ or $\mathfrak{R}$ of Theorem 2, Chapter 1, which in turn determines a constant b^*, ($|x - x^*| \leq b^*$, which is analogous to $\mathfrak{R}$ of Theorem 2, Chapter 1) to which the Peano existence theorem is applicable. Consequently, for each point $(y^*_1, \ldots, y^*_n, x^*)$ of this region we have a solution $y_j = \varphi^*_j(x)$ of Equations (24) for which Equation (22) holds. Since Equation (24) holds with y_j replaced by $\varphi^*_j(x)$ and y'_j replaced by $d\varphi^*_j/dx$; Equation (25) implies Equation (23) and this completes the proof of the theorem.

3. Continuation of the implicit function theorem

3.1 In future applications we shall need the following continuation of the implicit function theorem.

Theorem 3. Under the basic hypothesis H1, H2, H3 and the additional hypothesis

H4. In the open region $\mathfrak{A}$, the $\partial F_i/\partial x_j$ exist and are continuous for $i = 1, \ldots, n$ and some fixed j;

then: $\partial \varphi_i/\partial x_j$ exists and is continuous for $|x_k - x_{k,0}| \leq b'$.

Proof. We shall prove this result for $i = 1$, that is, for $\dfrac{\partial \varphi_1}{\partial x_j}$. It is clear that no loss of generality will result.

Turning back to the proof of Lemma 3, we let

$$x_k'' = x_k' + \delta_{kj}\,\Delta x_j, \qquad k = 1, \ldots, m$$

where δ_{kj} is the Kronecker delta. For $r = 1$ Equation (20) becomes

$$\Delta\varphi_1 = \varphi_1(x_1'', \ldots, x_m'') - \varphi_1(x_1', \ldots, x_m') = -\frac{K_1}{K} \tag{26}$$

where K_1 is

$$K_1 = \begin{vmatrix} \Delta F_1 \dfrac{\partial F_1}{\partial z_2} \cdots \dfrac{\partial F_1}{\partial z_n} \\ \cdots\cdots\cdots \\ \Delta F_n \dfrac{\partial F_n}{\partial z_2} \cdots \dfrac{\partial F_n}{\partial z_n} \end{vmatrix}.$$

Divide both sides of Equation (26) by Δx_j. One may divide K_1 by Δx_j by dividing each ΔF_k by Δx_j. By hypothesis H4,

$$\lim_{\Delta x_j \to 0} \frac{\Delta F_k}{\Delta x_j} = \frac{\partial F_k}{\partial x_j}$$

and by the same argument used in the corollary to Lemma 3, $K_1/\Delta x_j$ approaches

$$\begin{vmatrix} \dfrac{\partial F_1}{\partial x_j} \dfrac{\partial F_1}{\partial z_2} \cdots \dfrac{\partial F_1}{\partial z_n} \\ \cdots\cdots\cdots \\ \dfrac{\partial F_n}{\partial x_j} \dfrac{\partial F_n}{\partial z_2} \cdots \dfrac{\partial F_n}{\partial z_n} \end{vmatrix} = \frac{\partial(F_1, \ldots, F_n)}{\partial(x_j, z_2, \ldots, z_n)} \tag{27}$$

as $\Delta x_j \to 0$. By the corollary itself, K approaches

$$\begin{vmatrix} \frac{\partial F_1}{\partial z_1} & \frac{\partial F_1}{\partial z_2} & \cdots & \frac{\partial F_1}{\partial z_n} \\ \cdot & \cdot & \cdot & \cdot \\ \frac{\partial F_n}{\partial z_1} & \frac{\partial F_n}{\partial z_2} & \cdots & \frac{\partial F_n}{\partial z_n} \end{vmatrix} = \frac{\partial(F_1, \ldots, F_n)}{\partial(z_1, \ldots, z_n)} \tag{28}$$

as $\Delta x_j \to 0$. Equations (27) and (28) are, of course, evaluated at $(z_1', \ldots, z_n', x_1', \ldots, x_m')$. These two equations imply that $\partial\varphi_1/\partial x_j$ exists and

$$\frac{\partial\varphi_1}{\partial x_j} = -\frac{\dfrac{\partial(F_1, \ldots, F_n)}{\partial(x_j, z_2, \ldots, z_n)}}{\dfrac{\partial(F_1, \ldots, F_n)}{\partial(z_1, \ldots, z_n)}} \tag{29}$$

for $| x_i' - x_{i,0} | \leq b'$.

By H2 and H4, $\partial\varphi_1/\partial x_j$ is continuous.

Clearly, as remarked earlier, the same argument holds for $\partial\varphi_i/\partial x_j$, $i = 2, \ldots, n$. Also, if we assume that $\partial F_i/\partial x_j$ exists for $i = 1, \ldots, n$, $j = 1, \ldots, n$ then the $\partial\varphi_i/\partial x_j$ exist and are continuous for $i, j = 1, \ldots, n$ in $| x_j - x_{j,0} | \leq b'$, $j = 1, \ldots, n$.

3.2 While the right hand side of Equation (29) contains the φ_j, it does not contain any derivatives of these functions. Thus if we can differentiate the F_i's a second time, we may differentiate $\partial\varphi_i/\partial x_j$ again. In general, the explicit formula of Equation (29) may be used to establish immediately the following corollary to Theorem 3.

Corollary. If in addition to the hypotheses H1, H2, H3, H4 one has for any given set of non-negative integers $r_1, \ldots, r_m$ with $r_1 + r_2 + \ldots + r_m = r$ that for every combination of integers $r_1', \ldots, r_m', s_1, \ldots, s_n$ with $r_j' \leq r_j$ and $r_1' + \ldots + r_m' + s_1 + \ldots + s_n = r$ that

$$\frac{\partial^r F_i}{\partial z_1^{s_1} \ldots \partial z_n^{s_n} \partial x_1^{r_1'} \ldots \partial x_m^{r_m'}}$$

exists and is continuous on $\mathfrak{A}$, then

$$\frac{\partial^r \varphi_i}{\partial x_1^{r_1} \ldots \partial x_m^{r_m}}$$

exists and is continuous for $| x_j - x_{j,0} | \leq b'$.

4. Existence theorem in the general case

One more point remains to be considered in order to complete our discussion of elementary general existence theorems. We considered in our previous work systems of differential equations of the form

$$F_i(y_1', \ldots, y_n', y_1, \ldots, y_n, x) = 0, \qquad i = 1, \ldots, n, \tag{30}$$

that is, systems in which the highest derivative appearing was the first. We now consider differential equations of higher order, that is, systems of the form

$$F_i(z_1^{(d_1)}, z_1^{(d_1-1)}, \ldots, z_1, z_2^{(d_2)}, z_2^{(d_2-1)}, \ldots, z_2, \ldots, z_r^{(d_r)}, z_r^{(d_r-1)}, \ldots, z_r, x) = 0,$$
$$i = 1, \ldots, r \tag{31}$$

where z_j, $j = 1, \ldots, r$ are the unknown functions whose derivatives up to the order d_j respectively appear.

More precisely, we shall show in Theorem 4 below that under certain conditions, a higher order system of r equations in r unknowns, Equation (31), is equivalent to a first order system of n equations in n unknowns, Equation (30), where $n = \sum_{j=1}^{r} d_j$ and where the first order system satisfies the hypothesis of the Peano existence theorem in the implicit case. (By "equivalent" we mean that every solution of one system corresponds to a solution of the other, and conversely.) Thus, by applying the Peano existence theorem in the implicit case (Section 2) we obtain a solution of Equation (30) and hence of Equation (31).

Theorem 4. Hypothesis. (i) Let

$$F_i(z_1^{(d_1)}, \ldots, z_1, z_2^{(d_2)}, \ldots, z_2, \ldots, z_r^{(d_r)}, \ldots, z_r, x) = 0, \tag{32}$$
$$i = 1, \ldots, r$$

be r real valued functions of the $d_1 + \ldots + d_r + r + 1$ real variables $z_1^{(d_1)}, \ldots, z_1', z_1, z_2^{(d_2)}, \ldots, z_2', z_2, \ldots, z_r^{(d_r)}, \ldots, z_r', z_r, x$ defined and continuous on an open region $\mathfrak{A}$ of $(d_1+\ldots+d_r+r+1)$-dimensional euclidean space.

(ii) Let $\dfrac{\partial F_i}{\partial z_j^{(k)}}$ exist and be continuous on $\mathfrak{A}$ for $i, j = 1, \ldots, r$, and $k \leq d_j$.

(iii) Let $(z_{1,0}^{(d_1)}, \ldots, z_{1,0}, z_{2,0}^{(d_2)}, \ldots, z_{2,0}, \ldots, z_{r,0}^{(d_r)}, \ldots, z_{r,0}, x_0)$ be a point in $\mathfrak{A}$ such that

$$F_i(z_{1,0}^{(d_1)}, \ldots, z_{1,0}, z_{2,0}^{(d_2)}, \ldots, z_{2,0}, \ldots, z_{r,0}^{(d_r)}, \ldots, z_{r,0}, x_0) = 0, \qquad i = 1, \ldots, r$$

and such that the Jacobian J

$$J = \frac{\partial(F_1, \ldots, F_r)}{\partial(z_1^{(d_1)}, \ldots, z_r^{(d_r)})} \neq 0$$

at this point.

Conclusion. The given system, Equation (32) is equivalent (in the above defined sense) to a system of n differential equations of the type

$$F_i(y_1', \ldots, y_n', y_1, \ldots, y_n, x) = 0 \qquad i = 1, \ldots, n$$

where $n = \sum_{j=1}^{r} d_j$ and this system of equations satisfies the hypothesis of the Peano existence theorem in the implicit case.

Proof. Introduce $d_1 + \ldots + d_r$ new variables y_j by the equations

$$\begin{aligned}
y_1 &= z_1 \\
y_2 &= z_1' \\
&\cdots \\
y_{d_1} &= z_1^{(d_1-1)} \\
y_{d_1+1} &= z_2 \\
y_{d_1+2} &= z_2' \\
&\cdots \\
y_{d_1+d_2} &= z_2^{(d_2-1)} \\
&\cdots \\
&\cdots \\
y_{d_1+\ldots+d_{r-1}+1} &= z_r \\
y_{d_1+\ldots+d_{r-1}+2} &= z_r' \\
&\cdots \\
y_{d_1+\ldots+d_r} &= z_r^{(d_r-1)}.
\end{aligned}$$

The system of Equation (32) may thus be written,

$$
\begin{aligned}
&F_i(y'_{d_1}, y_{d_1}, y_{d_1-1}, \ldots, y_1, \quad y'_{d_1+d_2}, y_{d_1+d_2}, \ldots, y_{d_1+1}, \ldots,\\
&y'_{d_1+\cdots+d_r}, y_{d_1+\cdots+d_r}, \ldots, y_{d_1+\cdots+d_{r-1}+1}, x) = 0, \quad i = 1, \ldots, r\\
&F_{r+1} \equiv y'_1 - y_2 = 0\\
&F_{r+2} \equiv y'_2 - y_3 = 0\\
&\cdots\cdots\\
&F_{r+d_1-1} \equiv y'_{d_1-1} - y_{d_1} = 0\\
&F_{r+d_1+1} \equiv y'_{d_1+1} - y_{d_1+2} = 0\\
&F_{r+d_1+2} \equiv y'_{d_1+2} - y_{d_1+3} = 0\\
&\cdots\cdots\\
&F_{r+d_1+d_2-1} \equiv y'_{d_1+d_2-1} - y_{d_1+d_2} = 0\\
&\cdots\cdots\\
&\cdots\cdots\\
&F_{r+d_1+\cdots+d_{r-1}+1} \equiv y'_{d_1+\cdots+d_{r-1}+1} - y_{d_1+\cdots+d_{r-1}+2} = 0\\
&F_{r+d_1+\cdots+d_{r-1}+2} \equiv y'_{d_1+\cdots+d_{r-1}+2} - y_{d_1+\cdots+d_{r-1}+3} = 0\\
&\cdots\cdots\\
&F_{r+d_1+\cdots+d_r-1} \equiv y'_{d_1+\cdots+d_r-1} - y_{d_1+\cdots+d_r} = 0
\end{aligned}
\tag{33}
$$

which is a system of $r + (d_1 - 1) + (d_2 - 1) + \ldots + (d_r - 1) = \sum_{i=1}^{r} d_i = n$ differential equations.

We must show that the system of differential equations of Equation (33) satisfies the hypothesis of the Peano existence theorem and that it is equivalent to Equations (32).

By (i) of the hypothesis the F_i are continuous functions of $d_1 + \ldots + d_r + r + 1$ variables in $\mathfrak{A}$. Clearly, the remaining equations are also continuous functions of these variables in any region (and hence in $\mathfrak{A}$). By (ii) of the hypothesis $\partial F_i/\partial y'_{d_1}$, $\partial F_i/\partial y'_{d_1+d_2}$, $\ldots$, $\partial F_i/\partial y'_{d_1+\cdots+d_r}$ exist and are continuous on $\mathfrak{A}$. Clearly, the remaining equations have this property. By (iii) of the hypothesis the point $(y'_{d_1,0}, \ldots, y_{1,0}, y'_{d_1+d_2,0}, \ldots, y_{d_1+1,0}, \ldots, y'_{d_1+\cdots+d_r,0}, \ldots, y_{d_1+\cdots+d_{r-1}+1,0}, x_0)$ satisfies the F_i equations. By an explicit construction, the Jacobian of the system of Equation (33) is

$$\underbrace{\left|\begin{array}{cccc|cccc} \frac{\partial F_1}{\partial y'_{d_1}} & \frac{\partial F_1}{\partial y'_{d_1+d_2}} & \cdots & \frac{\partial F_1}{\partial y'_{d_1+\cdots+d_r}} & \theta_{11} & \theta_{12} & \cdots & \theta_{1,n-r} \\ \frac{\partial F_2}{\partial y'_{d_1}} & \frac{\partial F_2}{\partial y'_{d_1+d_2}} & \cdots & \frac{\partial F_2}{\partial y'_{d_1+\cdots+d_r}} & \theta_{21} & \theta_{22} & \cdots & \theta_{2,n-r} \\ \cdot & \cdot & \cdot & \cdot & \cdot & \cdot & \cdot & \cdot \\ \frac{\partial F_r}{\partial y'_{d_1}} & \frac{\partial F_r}{\partial y'_{d_1+d_2}} & \cdots & \frac{\partial F_r}{\partial y'_{d_1+\cdots+d_r}} & \theta_{r1} & \theta_{r2} & \cdots & \theta_{r,n-r} \\ \hline 0 & 0 & \cdots & 0 & 1 & 0 & \cdots & 0 \\ 0 & 0 & \cdots & 0 & 0 & 1 & \cdots & 0 \\ \cdot & \cdot & \cdot & \cdot & \cdot & \cdot & \cdot & \cdot \\ 0 & 0 & \cdots & 0 & 0 & 0 & \cdots & 1 \end{array}\right|}_{r \qquad\qquad\qquad n-r} \begin{array}{l} \left.\right\} r \\ \left.\right\} n-r \end{array}$$

where the θ_{ij}'s are the partials of the F_i's with respect to the various other y'_k's. This Jacobian is precisely

$$\frac{\partial(F_1, \ldots, F_r)}{\partial(y'_{d_1}, y'_{d_1+d_2}, \ldots, y'_{d_1+\cdots+d_r})}$$

which is unequal to zero by (iii) of the hypothesis.

We therefore see that the hypothesis of the implicit function theorem is true for Equations (33). By our construction of the y'_j the solutions of Equations (33) and (31) are identical.

With this proof, all the elementary existence theory is completed.

CHAPTER 3

Uniqueness Theorems

1. Lipschitz condition

1.1 The continuity of a function, or a system of functions, is not sufficient to guarantee the *uniqueness* of a solution through a specified point. For example, consider

$$f(y, x) = y^{2/3}$$

which is continuous for every value of x and y. However, the solution of

$$y' = y^{2/3} \tag{1}$$

is not unique when the initial point is chosen on the x-axis.

Suppose we choose the initial point as (c, 0). Clearly, $y \equiv 0$ is a solution of Equation (1) which passes through this point. Also, by an elementary integration

$$y = \frac{1}{27}(x - c)^3$$

is a second solution which satisfies Equation (1) and passes through the point (c, 0). We see, therefore, that continuity of f(y, x) is not sufficient to establish uniqueness.

Uniqueness is an important practical property. The objective of most physical investigations is to obtain differential equations governing the change in a system. It is desirable that the equations obtained with the initial state uniquely specify the history of the system. This means that the solutions of the set of differential equations corresponding to a given set of initial conditions should be unique. Otherwise, the analysis given is not adequate to specify the behavior of the system.

1.2 The above example shows that a further condition is necessary if we wish to have uniqueness. A weak such *sufficiency* condition is a *Lipschitz condition,* which we now define.

Definition. If f(y, x) is a real valued function of the two real variables y and x defined and continuous in a convex open region U of two-dimensional euclidean space, then f is said to satisfy a *Lipschitz condition* if there exists a constant L such that for any pair of point (y_1, x) and (y_2, x) in $\mathfrak{A}$,

$$|f(y_1, x) - f(y_2, x)| \leq L |y_1 - y_2|.$$

So, a Lipschitz condition is a strong uniformity condition. It can be interpreted to mean that when x is fixed, the resulting function of y has bounded difference quotients, and the bound, L, is independent of x.

1.3 Continuity in y and x does not imply a Lipschitz condition. As an example, consider again the function $f(y, x) = y^{2/3}$, in some region $\mathfrak{A}$ containing the origin. Clearly, $y^{2/3}$ is jointly continuous in y and x. However, if we set $y_1 = 0$, $y_2 = y$, the condition

$$|y^{2/3}| \leq L |y| \tag{2}$$

is not satisfied for all $(y, x) \varepsilon \mathfrak{A}$. For suppose such an L existed. Let $y = 1/(2L)^3$. Then Equation (2) implies

$$\frac{1}{4L^2} \leq \frac{L}{(2L)^3}$$

or

$$1 \leq \tfrac{1}{2},$$

which is absurd.

1.4 On the other hand, a Lipschitz condition implies continuity in y *alone*. For suppose f(y, x) satisfies a Lipschitz condition in a convex open region U. We know that f(y, x) is continuous in y alone if, given an $\epsilon > 0$ and two points $(y_1, x), (y_2, x) \varepsilon \mathfrak{A}$, there exists a $\delta > 0$ such that

$$|f(y_1, x) - f(y_2, x)| < \varepsilon$$

whenever $|y_1 - y_2| < \delta$. Now let f(y, x) satisfy a Lipschitz condition in $\mathfrak{A}$ with constant L. Given $\varepsilon > 0$, let $\delta = \frac{\varepsilon}{L}$.

1.5 The existence and boundedness of the partial derivative $\partial f/\partial y$ in $\mathfrak{A}$ are sufficient to guarantee uniqueness, that is, they imply a Lipschitz condition.

For suppose $\partial f/\partial y$ is bounded on $\mathfrak{A}$. Then there exists a constant L such that

$$\left|\frac{\partial f}{\partial y}\right| \leq L. \tag{3}$$

Let (y_1, x) and (y_2, x) be any two points in $\mathfrak{A}$. Then by the Law of the Mean

$$| f(y_1, x) - f(y_2, x) | = \left|\frac{\partial f}{\partial y}(\bar{y}, x)\right| \cdot | y_1 - y_2 |$$

where $\bar{y}$ lies between y_1 and y_2. By Equation (3), the above equation becomes

$$| f(y_1, x) - f(y_2, x) | \leq L \, | y_1 - y_2 |,$$

which is a Lipschitz condition.

1.6 Although the existence and boundedness of the partial derivative is not *necessary* for a Lipschitz condition (so a Lipschitz condition is weaker, that is, more general), in most cases it is unlikely that a Lipschitz condition will be satisfied when the partial derivative condition is not satisfied. Hence, if this latter condition is not satisfied, the usual procedure is first to try to prove that a solution is not unique.

2. The basic uniqueness theorem

2.1 The fundamental uniqueness theorem for one differential equation is as follows.

Theorem 1. Hypothesis. Let $f(y, x)$ be a real valued function of the two real variables y, x defined and continuous on a convex open region U of two-dimensional euclidean space. Let $f(y, x)$ satisfy a Lipschitz condition on $\mathfrak{A}$.

Conclusion. For every point (y_0, x_0) of $\mathfrak{A}$ we can find a $b > 0$ and a function $y(x)$ which has a continuous first derivative in the neighborhood $\mathfrak{N}$, $| x - x_0 | \leq b$ such that

(i) $dy/dx \equiv f(y(x), x)$

(ii) $y(x_0) = y_0$

(iii) $y(x)$ is the only function having these properties.

We note that if the sentence: "Let $f(y, x)$ satisfy a Lipschitz condition on $\mathfrak{A}$." is deleted from the hypothesis, and (iii) is deleted from the conclusion, Theorem 1 becomes identical with the basic existence theorem of Chapter 1.

2.2 From Theorem 1 of Chapter 1 we know that a y(x) exists which satisfies $y' = f(y, x)$ and passes through the point (y_0, x_0). Now suppose that for $|x - x_0| \leq b$ two solutions y(x) and z(x) which pass through the point (y_0, x_0) exist. It follows from Equation (2) of Section 1.2 of Chapter 1 that

$$y(x) = y_0 + \int_{x_0}^{x} f(y(t), t)\, dt$$

and

$$z(x) = y_0 + \int_{x_0}^{x} f(z(t), t)\, dt.$$

If we substract these two equations and take absolute values,

$$|y(x) - z(x)| = \left| \int_{x_0}^{x} [f(y(t), t) - f(z(t), t)]\, dt \right|$$

$$\leq \left| \int_{x_0}^{x} | f(y, t) - f(z, t) |\, dt \right|.$$

Applying the Lipschitz condition, this becomes

$$|y(x) - z(x)| \leq \left| \int_{x_0}^{x} L\, |y - z|\, dt \right| = L \left| \int_{x_0}^{x} |y(t) - z(t)|\, dt \right|. \quad (4)$$

Now suppose $x \geq x_0$. (A similar proof holds for $x < x_0$.) Then we may remove the absolute values on the outside of the integral sign and write

$$|y(x) - z(x)| \leq L \int_{x_0}^{x} |y(t) - z(t)|\, dt. \quad (5)$$

Since y(x) and z(x) have continuous derivatives for $x \varepsilon \mathfrak{R}$ by the Peano existence theorem, they are continuous and hence bounded on the interval $|x - x_0| \leq b$. Thus there exists a constant D such that

$$|y(x)| \leq D, \quad |z(x)| \leq D \quad \text{for} \quad x \varepsilon \mathfrak{R}.$$

Therefore,

$$|y(x) - z(x)| \leq 2\,D \quad \text{for} \quad x \varepsilon \mathfrak{R}. \quad (6)$$

Placing Equation (6) in Equation (5) yields

$$|y(x) - z(x)| \leq L(2\,D)\,(x - x_0), \quad (7)$$

and placing Equation (7) in Equation (5) yields

$$|y(x) - z(x)| \leq L^2(2D) \int_{x_0}^{x} (t - x_0)\, dt = L^2(2D)\, \frac{(x - x_0)^2}{2!}. \quad (8)$$

By induction we conclude

$$| y(x) - z(x) | \leq L^k(2D) \frac{(x - x_0)^k}{k!}, \quad x \varepsilon \mathfrak{N}. \tag{9}$$

For suppose

$$| y(x) - z(x) | \leq L^{k-1}(2D) \frac{(x - x_0)^{k-1}}{(k-1)!}.$$

Placing this in Equation (5) yields

$$| y(x) - z(x) | \leq L(L^{k-1}) (2D) \int_{x_0}^{x} \frac{(t - x_0)^{k-1}}{(k-1)!} dt,$$

which upon integrating yields Equation (9).

We assert that

$$\lim_{k \to \infty} L^k(2D) \frac{(x - x_0)^k}{k!} = 0.$$

This can easily be proved directly, but an even simpler attack is to notice that

$$\frac{L^k(x - x_0)^k}{k!} \leq \frac{L^k b^k}{k!}$$

since $x \varepsilon \mathfrak{N}$. Now, $(L^k b^k)/k!$ is the $(k + 1)^{st}$ term in the power series expansion of e^{Lb}, and since this series converges,

$$\lim_{k \to \infty} \frac{L^k b^k}{k!} = 0.$$

Thus we conclude that $| y(x) - z(x) | \leq 0$ which implies $y(x) = z(x)$ for all $x \varepsilon \mathfrak{N}$.

2.3 In the above theorem we imposed certain conditions on a function f(y, x), *constructed* a neighborhood of the initial point x_0 and then showed that if two solutions satisfied the differential equation and had the same initial condition, they coincided on this neighborhood. A more general result is given in the corollary below where we assume that an interval (which is not necessarily to be thought of as small) is given upon which two solutions have a common point, and then prove these solutions coincide. This may be regarded as a result "in the large" as opposed to Theorem 1 which is a result "in the small."

Corollary. Hypothesis as in Theorem 1 and,

(i) y(x) and z(x) are two functions of x such that (y(x), x) and (z(x), x) remain in $\mathfrak{A}$ for an x interval $x_1 \leq x \leq x_2$,

(ii) $y(x)$ and $z(x)$ have a common initial point, that is $y(x') = z(x')$, $x_1 \leq x' \leq x_2$,

(iii) $y(x)$ and $z(x)$ are both solutions of $dy/dx = f(y, x)$ for $x \varepsilon [x_1, x_2]$.

Conclusion. $y(x)$ and $z(x)$ coincide on the interval $x_1 \leq x \leq x_2$.

Proof. Consider a range of x for which $x' \leq x \leq x_2$. (A similar argument holds for $x_1 \leq x \leq x'$.) Since $y(x)$ and $z(x)$ are solutions of the given differential equation on $[x_1, x_2]$ (and hence certainly on $[x', x_2]$) for which $y(x') = z(x')$, we may write

$$y(x) = y(x') + \int_{x'}^{x} f(y(t), t)\, dt$$

$$z(x) = y(x') + \int_{x'}^{x} f(z(t), t)\, dt.$$

Subtracting these two equations and taking absolute values,

$$| y(x) - z(x) | \leq \int_{x'}^{x} | f(y, t) - f(z, t) |\, dt.$$

Then the proof that $y(x) = z(x)$ is identical with that of Theorem 1 [cf. Equation (4)], where $D \geq | y(x) |$, $D \geq | z(x) |$ for $x \varepsilon [x_1, x_2]$ and b is replaced by $x_2 - x_1$, which of course is finite.

3. Uniqueness theorem for n unknown functions

3.1 For the case of functions $f(y_1, \ldots, y_n, x)$ in n dependent variables, defined and continuous on a convex open region $\mathfrak{A}$ of $(n + 1)$-dimensional euclidean space, the Lipschitz condition assumes the form

$$| f(y_1^*, \ldots, y_n^*, x) - f(y_1^+, \ldots, y_n^+, x) | \leq L \sum_{j=1}^{n} | y_j^* - y_j^+ |$$

for any two points $(y_1^*, \ldots, y_n^*, x)$ and $(y_1^+, \ldots, y_n^+, x)$ in $\mathfrak{A}$. As in the case of one variable, continuity in the $n + 1$ variables $y_1, \ldots, y_n, x$ does not imply a Lipschitz condition. However, a Lipschitz condition does imply joint continuity in the n *dependent* variables $y_1, \ldots, y_n$. Also, if the $\partial f/\partial y_j$ exist and are bounded on $\mathfrak{A}$, f satisfies a Lipschitz condition on $\mathfrak{A}$. (Cf. the proof of Theorem 3 of this chapter.)

3.2 We now state our uniqueness theorem. (Cf. Theorem 2 of Chapter 1.)

Theorem 2. Hypothesis. (i) Let $f_i(y_1, \ldots, y_n, x)$ be n real valued functions of the $n + 1$ real variables $y_1, \ldots, y_n, x$ defined and continuous on a convex open region U of $(n + 1)$-dimensional euclidean space.

(ii) (Lipschitz condition). There exists a constant L such that for every pair of points $(y_1^*, \ldots, y_n^*, x)$ and $(y_1^+, \ldots, y_n^+, x)$ in $\mathfrak{A}$ we have

$$| f_i(y_1^*, \ldots, y_n^*, x) - f_i(y_1^+, \ldots, y_n^+, x) | \leq L \sum_{j=1}^{n} | y_j^* - y_j^+ |, \qquad i = 1, \ldots, n.$$

Conclusion. For every point $(y_{1,0}, \ldots, y_{n,0}, x_0)$ of $\mathfrak{A}$ we can find a $b > 0$ and n functions $y_1(x), \ldots, y_n(x)$ which have continuous first derivatives in the neighborhood $\mathfrak{N}$, $| x - x_0 | \leq b$ such that

(i) $dy_i/dx = f_i(y_1(x), \ldots, y_n(x), x)$, $i = 1, \ldots, n$.

(ii) $y_i(x_0) = y_{i,0}$, $i = 1, \ldots, n$.

(iii) The $y_1(x), \ldots, y_n(x)$ are the only set of functions having these properties.

Compare this theorem with the Peano existence theorem of Section 2.1 of Chapter 1. Except for the Lipschitz condition in the hypothesis and (iii) of the conclusions, the theorems are identical. As in the case of Theorem 1 of this chapter our results are in the small; however, we have an analogous corollary "in the large." (Cf. Section 3.4.)

3.3 The proof of Theorem 2 parallels that for the case $n = 1$.

From Theorem 2 of Chapter 1 we know there exist n functions $y_1(x), \ldots, y_n(x)$ which satisfy $dy_i/dx = f_i(y_1, \ldots, y_n, x)$ and pass through the point x_0. Now suppose that for $| x - x_0 | \leq b$ two sets of solutions $y_1(x), \ldots, y_n(x)$ and $z_1(x), \ldots, z_n(x)$ exist which have the same initial conditions,

$$y_j(x_0) = z_j(x_0) = y_{j,0}, \qquad j = 1, \ldots, n.$$

It follows then that

$$y_j(x) = y_{j,0} + \int_{x_0}^{x} f_j(y_1(t), \ldots, y_n(t), t)\, dt, \qquad j = 1, \ldots, n$$

$$z_j(x) = y_{j,0} + \int_{x_0}^{x} f_j(z_1(t), \ldots, z_n(t), t)\, dt, \qquad j = 1, \ldots, n$$

for $x \in \mathfrak{R}$. Let us assume $x \geqq x_0$. (A similar proof holds for $x < x_0$.) Then if we subtract the above two equations for corresponding values of j and take absolute values,

$$|y_j - z_j| \leq \int_{x_0}^{x} |f_j(y_1, \ldots, y_n, t) - f_j(z_1, \ldots, z_n, t)|\, dt, \quad j = 1, \ldots, n.$$

And since the t is the same in both functions in the integrand, we may apply the Lipschitz condition of the hypothesis to obtain

$$|y_j - z_j| \leq \int_{x_0}^{x} L \sum_{k=1}^{n} |y_k - z_k|\, dt, \quad j = 1, \ldots, n.$$

If we now sum the above expressions over j,

$$\sum_{j=1}^{n} |y_j - z_j| \leq \int_{x_0}^{x} L\, n \sum_{k=1}^{n} |y_k - z_k|\, dt = nL \int_{x_0}^{x} \sum_{j=1}^{n} |y_j - z_j|\, dt. \tag{10}$$

As in the case $n = 1$, there exist D_j such that

$$|y_j(x)| \leq D_j, \qquad |z_j(x)| \leq D_j, \qquad j = 1, \ldots, n$$

for $x \in \mathfrak{R}$ and if we let

$$D = \sum_{j=1}^{n} D_j,$$

$$\sum_{j=1}^{n} |y_j(x) - z_j(x)| \leq 2D. \tag{11}$$

Placing Equation (11) in Equation (10) yields

$$\sum_{j=1}^{n} |y_j - z_j| \leq nL \int_{x_0}^{x} 2D\, dt = (nL)(2D)(x - x_0).$$

By an induction similar to that used in the proof of the last theorem,

$$\sum_{j=1}^{n} |y_j(x) - z_j(x)| \leq 2D(nL)^k \frac{(x - x_0)^k}{k!}.$$

As before

$$\lim_{k \to \infty} (nL)^k \frac{(x - x_0)^k}{k!} = 0$$

(since n is fixed). Therefore

$$\sum_{j=1}^{n} |y_j(x) - z_j(x)| = 0$$

which implies $| y_j(x) - z_j(x) | = 0$ for all j, or

$$y_j(x) = z_j(x), \quad j = 1, \ldots, n.$$

3.4 The corollary to Theorem 2 (which is a result in the large) is similar in scope and proof to the corollary to Theorem 1.

Corollary. Hypothesis as in Theorem 2 and

(i) $y_1(x), \ldots, y_n(x)$ and $z_1(x), \ldots, z_n(x)$ are two sets of functions such that $(y_1(x), \ldots, y_n(x), x)$ and $(z_1(x), \ldots, z_n(x), x)$ remain in $\mathfrak{A}$ for an x interval $x_1 \leq x \leq x_2$,

(ii) $y_1(x), \ldots, y_n(x)$ and $z_1(x), \ldots, z_n(x)$ have common initial points, that is, $y_j(x') = z_j(x')$, $j = 1, \ldots, n$ for $x_1 \leq x' \leq x_2$,

(iii) $y_1(x), \ldots, y_n(x)$ and $z_1(x), \ldots, z_n(x)$ are both solutions of $dy_i/dx = f_i(y_1, \ldots, y_n, x)$, $i = 1, \ldots, n$ for $x \in [x_1, x_2]$.

Conclusion. On the interval $x_1 \leq x \leq x_2$, $y_j(x) = z_j(x)$, $j = 1 \ldots, n$.

Proof. We are given that x' is in the interval $[x_1, x_2]$. Let $x' \leq x \leq x_2$. (A similar argument will hold for $x_1 \leq x \leq x'$.) Since $y_1(x), \ldots, y_n(x)$ and $z_1(x), \ldots, z_n(x)$ are solutions of the given system on $[x_1, x_2]$, and hence on $[x', x_2]$ for which $y_j(x') = z_j(x')$, $j = 1, \ldots, n$, it follows that

$$y_j = y_j(x') + \int_{x'}^{x} f_j(y_1(t), \ldots, y_n(t), t)\, dt, \quad j = 1, \ldots, n$$

$$z_j = y_j(x') + \int_{x'}^{x} f_j(z_1(t), \ldots, z_n(t), t)\, dt, \quad j = 1, \ldots, n.$$

Subtracting for corresponding values of j and taking absolute values, we obtain

$$| y_j - z_j | \leq \int_{x'}^{x} | f_j(y_1, \ldots, y_n, t) - f_j(z_1, \ldots, z_n, t) |\, dt, \quad j = 1, \ldots n.$$

The remainder of the proof is identical with that of Theorem 2.

4. Uniqueness in the implicit case

4.1 Now we wish to state and prove a theorem that will give us uniqueness in the implicit case. We are concerned with differential equations of the form

$$F_i(y_1', \ldots, y_n', y_1, \ldots, y_n, x) = 0. \tag{12}$$

With suitable hypotheses, in particular the existence and continuity of the partial derivatives $\partial F_i/\partial y'_j$ we can obtain a solution. (Cf. Theorem 2 of Chapter 2.) However, for uniqueness we must impose an additional condition. One such condition is to assume the existence and continuity of the $\partial F_i/\partial y_j$ partial derivatives. It will be seen that these imply a Lipschitz condition. (Cf. Sections 1.5 and 3.1 of this chapter.)

Theorem 3. Hypothesis. (i) Given the n functions

$$F_i(y'_1, \ldots, y'_n, y_1, \ldots, y_n, x) = 0, \quad i = 1, \ldots, n \tag{13}$$

of the $2n + 1$ real variables $y'_1, \ldots, y'_n, y_1, \ldots, y_n, x$ defined and continuous on a convex open region U of $(2n + 1)$-dimensional euclidean space.

(ii) $\partial F_i/\partial y'_j$ exist and are continuous on $\mathfrak{A}$ for $i, j = 1, \ldots, n$.

(iii) $\partial F_i/\partial y_j$ exist and are continuous on $\mathfrak{A}$ for $i, j = 1, \ldots, n$.

(iv) There exists a set of values $(y'_{1,0}, \ldots, y'_{n,0}, y_{1,0}, \ldots, y_{n,0}, x_0)$ in $\mathfrak{A}$ such that

$$F_i(y'_{1,0}, \ldots, y'_{n,0}, y_{1,0}, \ldots, y_{n,0}, x_0) = 0, \quad i = 1, \ldots, n$$

and

$$J = \frac{\partial(F_1, \ldots, F_n)}{\partial(y'_1, \ldots, y'_n)} \neq 0$$

at this point.

Conclusion. There exists a positive number b' such that for every set of $n + 1$ values $y^*_1, \ldots, y^*_n, x^*$ with $|y^*_j - y_{j,0}| < b'$ and $|x^* - x_0| < b'$ there exists a $b^* > 0$, $(b^* \leq b')$ and n functions $\varphi^*_1(x), \ldots, \varphi^*_n(x)$ with $y^*_j = \varphi^*_j(x^*)$ which have continuous first derivatives in the neighborhood $|x - x^*| \leq b^*$ such that

$$F_i\left(\frac{d\varphi^*_1}{dx}, \ldots, \frac{d\varphi^*_n}{dx}, \varphi^*_1, \ldots, \varphi^*_n, x\right) = 0, \quad i = 1, \ldots, n.$$

Furthermore, the $\varphi^*_i(x)$ are unique.

4.2 From Theorem 2 of Chapter 2 we know that there exist n functions $\varphi^*_1(x), \ldots, \varphi^*_n(x)$ which satisfy Equation (13) in the neighborhood $|x - x^*| \leq b^*$ and $\varphi^*_j(x^*) = y^*_j$, $j = 1, \ldots, n$. From the implicit function theorem we know that there exist n continuous functions f_i in the region $\mathfrak{B}$ defined by $|x - x_0| \leq b'$, $|y_j - y_{j,0}| \leq b'$ such that

$$y_j' = f_j(y_1, \ldots, y_n, x). \quad j = 1, \ldots, n.$$

Now if we can show that the f_j satisfy a Lipschitz condition in $\mathfrak{B}$, Theorem 2 will imply the uniqueness of the φ's in $|x - x^*| \leq b^*$.

By (iii) of the hypothesis the $\partial F_i/\partial y_j$ exist and are continuous and hence by Theorem 3 of Chapter 2 the $\partial f_i/\partial y_k$ exist and are continuous for $i, k = 1, \ldots, n$ in the neighborhood $\mathfrak{B}$. Since $\mathfrak{B}$ is closed, the $\partial f_i/\partial y_k$ are bounded, say

$$\left|\frac{\partial f_i}{\partial y_k}\right| \leq L_{ik}, \qquad i, k = 1, \ldots, n.$$

Let

$$L = \max_{i,k} L_{ik}. \tag{14}$$

Now if $(y_1^-, \ldots, y_n^-, x)$ and $(y_1^+, \ldots, y_n^+, x)$ are any two points in $\mathfrak{B}$, we can apply the Law of the Mean in n variables to obtain

$$f_i(y_1^-, \ldots, y_n^-, x) - f_i(y_1^+, \ldots, y_n^+, x)$$
$$= \sum_{j=1}^{n} \frac{\partial f_i}{\partial y_j}(\bar{y}_1, \ldots, \bar{y}_n, x)(y_j^- - y_j^+), \qquad i = 1, \ldots, n$$

where $\bar{y}_j$ lies between y_j^- and y_j^+. Taking absolute values,

$$|f_i(y_1^-, \ldots, y_n^-, x) - f_i(y_1^+, \ldots, y_n^+, x)| \leq \sum_{j=1}^{n} \left|\frac{\partial f_i}{\partial y_j}\right| |y_j^- - y_j^+|,$$
$$i = 1, \ldots, n.$$

By Equation (14)

$$|f_i(y_1^-, \ldots, y_n^-, x) - f_i(y_1^+, \ldots, y_n^+, x)| \leq L \sum_{j=1}^{n} |y_j^- - y_j^+|,$$
$$i = 1, \ldots, n$$

which is our Lipschitz condition.

4.3 A slightly more general uniqueness theorem than the one just proved can be stated. In the above theorem we assumed by (iii) of the hypothesis that the $\partial F_i/\partial y_j$ existed and were continuous. This was *sufficient* to establish a Lipschitz condition. However, if we delete this from the hypothesis, we still have, by the implicit function theorem, the existence of sets of functions

$$y_i' = f_i(y_1, \ldots, y_n, x)$$

defined in some region of the y and x variables. If we require the f_i functions to satisfy a Lipschitz condition, uniqueness will be established.

The theorem just proved, however, is more elegant and allows one to test the given F_i functions directly to see if the $\partial F_i/\partial y_j$ exist and are continuous (and thus insure uniqueness) without having to apply first the implicit function theorem. Of course, if it is known *a priori* that the f_i functions satisfy a Lipschitz condition, (iii) of the hypothesis becomes superfluous.

CHAPTER 4

The Picard Iterants

1. Introduction

1.1 When we have a Lipschitz condition on a region $\mathfrak{A}$, the initial condition (y_0, x_0) uniquely determines the solution of the equation $dy/dx = f(y, x)$ in a certain x interval; that is, corresponding to each point of $\mathfrak{A}$ we can find a unique solution valid on this interval. Hence we have a functional relationship involving initial conditions and solutions, that is, $y = y(x, y_0, x_0)$ where y_0 and x_0 may be regarded as parameters. Later we shall study the dependence of solutions not only on x but on the initial conditions also. We see that without uniqueness the solutions would not involve the parameters in a functional relationship, and we would not have such a theory. In other words, in order to study the dependence of solutions on initial conditions, we must have uniqueness.

The methods we shall develop will also enable us to consider differential equations of the type $y' = f(y, x, \lambda)$ where λ is a parameter. We shall want to study the dependence of solutions on the parameter λ,

$$y = y(x, \lambda)$$

as well as

$$y = y(x, y_0, x_0, \lambda).$$

Again, uniqueness is imperative. Clearly, our methods will also be applicable to differential equations in n dependent variables.

The importance of differential equations involving parameters cannot be overestimated, parameters being very important in practical problems. This dependence of solutions on parameters is one of the cardinal reasons for studying differential equations. It is mainly to provide a background necessary for the investigation of such properties that the studies of the present chapter are undertaken.

1.2 We know that the differential equation

$$\frac{dy}{dx} = f(y, x) \tag{1}$$

with the initial condition (y_0, x_0) and certain appropriate assumptions is equivalent to the integral equation

$$y(x) = y_0 + \int_{x_0}^{x} f(y(t), t)\, dt. \tag{2}$$

Now this suggests that we can start with a function $y^{(0)}(t)$ and repeatedly substitute in the right hand side of Equation (2) to obtain a sequence of functions,

$$y^{(1)}(x) = y_0 + \int_{x_0}^{x} f(y^{(0)}(t), t)\, dt \text{ [1]}$$

$$y^{(2)}(x) = y_0 + \int_{x_0}^{x} f(y^{(1)}(t), t)\, dt$$

$$\cdots\cdots\cdots\cdots$$

$$y^{(k)}(x) = y_0 + \int_{x_0}^{x} f(y^{(k-1)}(t), t)\, dt$$

$$\cdots\cdots\cdots\cdots$$

$$\cdots\cdots\cdots\cdots$$

Now if the sequence $\{y^{(k)}(x)\}$ converges uniformly to a function $y(x)$ then we may conclude, as in Lemma 4 of Chapter 1 that $y(x)$ will be a solution of the integral equation, Equation (2), and hence by Lemma 2 of Chapter 1, a solution of the original differential equation, Equation (1). This offers us not only a method of proof of the existence of the solution of the original differential equation, but also an *explicit* method for expressing the solution as a limit of known functions. This explicit representation allows us to obtain many properties of the solutions. In particular, it allows us to investigate the dependence of solutions on initial conditions and parameters.

1.3 The functions $\{y^{(k)}(x)\}$ discussed above are called, when precisely defined, the *Picard iterants.* To show the significance of these iterants, we shall prove (independently of the above) the

[1] In $y^{(1)}, y^{(2)}, \ldots, y^{(k)}, \ldots$ the $1, 2, \ldots, k, \ldots$ are superscripts and not derivatives.

existence and uniqueness of a solution of the differential equation $dy/dx = f(y, x)$ by the explicit construction of the Picard iterants, $\{y^{(k)}(x)\}$. In order to obtain the uniform convergence of the iterants we must impose a Lipschitz condition on $f(y, x)$. Hence our existence proof will not be as general as the Peano construction, and the method of proof will be radically different. Also our uniqueness proof using the Picard iterants will differ from the uniqueness proof given in the last chapter.

Theorem 1. Let $f(y, x)$ be a real valued function of the two real variables y, x defined and continuous on a convex open region $\mathfrak{A}$ of two-dimensional euclidean space and satisfying a Lipschitz condition in this region. Then for every point (y_0, x_0) of $\mathfrak{A}$ we find a $b > 0$ and a function $y(x)$ which has a continuous first derivative in the neighborhood $\mathfrak{N}$, $|x - x_0| \leq b$ such that

$$\frac{dy}{dx} = f(y(x), x) \tag{3}$$

and $y(x_0) = y_0$ in the neighborhood $\mathfrak{N}$. Furthermore, $y(x)$ is unique.

Proof. Since $\mathfrak{A}$ is open there exists a rectangular neighborhood $\mathfrak{R}$ of (y_0, x_0), say

$$|x - x_0| \leq a, \qquad |y - y_0| \leq a$$

which lies entirely within $\mathfrak{A}$. As in the Peano case (Lemma 2 of Chapter 1), Equation (3) is equivalent to an integral equation

$$y(x) = y_0 + \int_{x_0}^{x} f(y(t), t)\, dt. \tag{4}$$

We shall now explicitly construct a sequence of functions $\{y^{(k)}(x)\}$ (the Picard iterants) which converge uniformly to a solution of Equation (4) and hence to a solution of Equation (3).

Define

$$y^{(0)}(x) = y_0$$
$$y^{(1)}(x) = y^{(0)} + \int_{x_0}^{x} f(y^{(0)}(t), t)\, dt \tag{5}$$

and recursively

$$y^{(k+1)}(x) = y^{(0)} + \int_{x_0}^{x} f(y^{(k)}(t), t)\, dt.$$

Clearly $(y^{(0)}, x) \in \mathfrak{A}$ and since f is continuous, $f(y^{(0)}, t)$ is continuous and hence $y^{(1)}(x)$ is continuous. Now consider

$$y^{(2)}(x) = y^{(0)} + \int_{x_0}^{x} f(y^{(1)}(t), t)\, dt.$$

Since $y^{(1)}$ is continuous, $y^{(2)}$ is continuous. However, we do not know that $y^{(2)}$ exists at all unless we can prove that $(y^{(1)}(t), t) \in \mathfrak{A}$, that is, the region of definition of f. Hence our first task is to show that, in general, $(y^{(k)}, x) \in \mathfrak{A}$ and hence that the $y^{(k)}(x)$ are well defined.

In order to do this we proceed as in the Peano theorem. Since $f(y, x)$ is continuous in the closed rectangle $\mathfrak{R}$ it is bounded, say

$$|f(y, x)| \leq M \quad \text{on} \quad \mathfrak{R}.$$

Now let

$$b = \min\left(a, \frac{a}{M}\right)$$

and call $\mathfrak{R}'$ the region defined by $|x - x_0| \leq b$, $|y - y_0| \leq a$. From Equation (5)

$$|y^{(1)} - y^{(0)}| = \left|\int_{x_0}^{x} f(y^{(0)}(t), t)\, dt\right| \leq M\,|x - x_0| \leq Mb \leq a$$

for $|x - x_0| \leq b$. [Since $(y^{(0)}, x) \in \mathfrak{R}'$, $|f(y^{(0)}, x)| \leq M$.] The proof now proceeds by induction.

Assume

$$|y^{(k)} - y^{(0)}| \leq a$$

for $x \in \mathfrak{R}$. This implies $(y^{(k)}, x) \in \mathfrak{R}'$ and hence $|f(y^{(k)}, x)| \leq M$. Thus

$$|y^{(k+1)} - y^{(0)}| \leq \left|\int_{x_0}^{x} |f(y^{(k)}, t)|\, dt\right| \leq M\,|x - x_0| \leq a$$

and $(y^{(k+1)}, x) \in \mathfrak{R}'$.

The existence of the iterants $y^{(k)}(x)$ has thus been established.

We note in passing that if the function $f(y, x)$ remains continuous for all $x \in [x_0 - a,\ x_0 + a]$ and all y, $y \in (-\infty, +\infty)$, then it is unnecessary to introduce the "cut down neighborhood" defined by $\mathfrak{R}'$ since for any x, y is always in $\mathfrak{A}$.

Now assume $x \geq x_0$. (A similar proof holds for $x < x_0$.) From Equation (5)

$$|y^{(1)} - y^{(0)}| = \left|\int_{x_0}^{x} f(y^{(0)}, t)\, dt\right| \leq M(x - x_0)$$

for $x \in \mathfrak{R}$. Subtract $y^{(1)}$ from $y^{(2)}$ and take absolute values.

$$|y^{(2)} - y^{(1)}| = \left|\int_{x_0}^{x} [f(y^{(1)}, t) - f(y^{(0)}, t)]\, dt\right|.$$

Applying the Lipschitz condition of the hypothesis,

$$|y^{(2)} - y^{(1)}| \leq L \int_{x_0}^{x} |y^{(1)} - y^{(0)}|\, dt \leq LM \int_{x_0}^{x} (t - x_0)\, dt = LM \frac{(x - x_0)^2}{2!}$$

for $|x - x_0| \leq b$ where L is the constant of the Lipschitz condition.

We shall now prove inductively that

$$|y^{(k)} - y^{(k-1)}| \leq ML^{k-1} \frac{(x - x_0)^k}{k!} \tag{6}$$

for all k. Assume that Equation (6) is valid. Then

$$\begin{aligned} |y^{(k+1)} - y^{(k)}| &= \left| \int_{x_0}^{x} [f(y^{(k)}, t) - f(y^{(k-1)}, t)]\, dt \right| \\ &\leq L \int_{x_0}^{x} |y^{(k)} - y^{(k-1)}|\, dt \leq ML^k \int_{x_0}^{x} \frac{(t - x_0)^k}{k!}\, dt \\ &= ML^k \frac{(x - x_0)^{k+1}}{(k+1)!}. \end{aligned}$$

Consider now the series

$$y = y^{(0)} + (y^{(1)} - y^{(0)}) + (y^{(2)} - y^{(1)}) + \ldots + (y^{(k)} - y^{(k-1)}) + \ldots. \tag{7}$$

Each term $(y^{(k)} - y^{(k-1)})$ is dominated by the expression

$$ML^{k-1} \frac{(x - x_0)^k}{k!}$$

which in turn is dominated by the constant $ML^{k-1} \frac{b^k}{k!}$. Now

$$|y_0|_{max} + Mb + \frac{MLb^2}{2!} + \frac{ML^2b^3}{3!} + \ldots \tag{8}$$

is a series of constant terms which dominates Equation (7). Equation (8) is an absolutely convergent power series, in fact it is precisely equal to

$$|y_0|_{max} + \frac{M}{L} M(e^{Lb} - 1).$$

Hence by the Weierstrass M-test, Equation (7) converges uniformly and absolutely. Since the $y^{(k)}(x)$ are continuous, $y(x)$ is continuous for $|x - x_0| \leq b$.

As in Lemma 4 of Chapter 1

$$\lim_{k \to \infty} \int_{x_0}^{x} f(y^{(k)}, t)\, dt = \int_{x_0}^{x} f(y,t)\, dt$$

and hence

$$y(x) = y_0 + \int_{x_0}^{x} f(y(t), t)\, dt, \qquad x \varepsilon \mathfrak{R}.$$

Lemma 2 of Chapter 1 completes the existence proof.

To establish uniqueness, assume that, in addition to the solution $y(x)$ obtained above, another solution $Y(x)$ defined for $x \varepsilon \mathfrak{R}$ exists and assumes the same initial value, that is $Y(x_0) = y_0$.

Since $Y(x)$ is a solution of $dy/dx = f(y, x)$ we may write [Equation (4)],

$$Y(x) = y_0 + \int_{x_0}^{x} f(Y(t), t)\, dt.$$

Now the $(k + 1)$ *st* Picard iterant leading to the function $y(x)$ is given by

$$y^{(k+1)}(x) = y_0 + \int_{x_0}^{x} f(y^{(k)}(t), t)\, dt.$$

Subtracting these two equations and taking absolute values,

$$| Y(x) - y^{(k+1)}(x) | \leq \left| \int_{x_0}^{x} | f(Y, t) - f(y^{(k)}, t) |\, dt \right|.$$

We shall show that $| Y(x) - y^{(k+1)}(x) |$ converges uniformly to zero as k approaches infinity and hence

$$Y(x) \equiv y(x),$$

which establishes the uniqueness.

By the Lipschitz condition

$$| Y(x) - y^{(k+1)}(x) | \leq \left| \int_{x_0}^{x} | f(Y,t) - f(y^{(k)},t) |\, dt \right|$$
$$\leq L \left| \int_{x_0}^{x} |Y - y^{(k)}| dt \right|.$$

Assume $x \geq x_0$. (A similar argument holds for $x < x_0$.) Then the above inequality becomes, for $k = 0$,

$$| Y(x) - y^{(1)}(x) | \leq L \int_{x_0}^{x} | Y - y_0 |\, dx \leq La(x - x_0)$$

[since $| Y - y_0 | \leq a$, (Y, x) being a point of $\mathfrak{R}'$ for $x \varepsilon \mathfrak{R}$].

We shall prove by induction that

$$| Y(x) - y^{(k)}(x) | \leq a L^k \frac{(x - x_0)^k}{k!}$$

For, assume the truth of the above inequality for k = k. Then for k + 1

$$| Y(x) - y^{(k+1)}(x)| \leq L \int_{x_0}^{x} | Y - y^{(k)} | dt \leq aL \int_{x_0}^{x} L^k \frac{(t - x_0)^k}{k!} dt$$

$$= aL^{k+1} \frac{(x - x_0)^{k+1}}{(k + 1)!}.$$

Now, $| x - x_0 | \leq b$. Hence

$$| Y(x) - y^{(k+1)}(x) | \leq \frac{a(Lb)^{k+1}}{(k + 1)!} \quad \text{(independent of x)}.$$

Clearly

$$\lim_{k \to \infty} \frac{a(Lb)^{k+1}}{(k + 1)!} = 0$$

and hence

$$Y(x) = \lim_{k \to \infty} y^{(k)}(x) = y(x).$$

2. The Picard iterants

2.1 Our main concern is with systems of differential equations of the form

$$\frac{dy_i}{dx} = f_i(y_1, \ldots, y_n, x), \qquad i = 1, \ldots, n \tag{9}$$

or

$$\frac{dy_i}{dx} = f_i(y_1, \ldots, y_n, x, \lambda_1, \ldots, \lambda_r), \qquad i = 1, \ldots, n$$

which satisfy a Lipschitz condition. We shall first define the Picard iterants in the case of Equation (9) and prove some of their basic properties.

We recall that the functions $\{y^{(k)}(x)\}$ obtained in Theorem 1 were called Picard iterants. Naturally there exist corresponding sequences of functions in the case of n dependent variables which we must define precisely. However, as in the case of n = 1 there is a question as to the possibility of forming these sequences, since in order to carry out the kth step it is necessary to apply the functions f_i to the results of the $(k - 1)^{st}$ step. In the case of functions of one variable, this required that $(y^{(k)}(x), x)$ be in the region $\mathfrak{R}'$, or, more practically, in the region $\mathfrak{A}$ for each value of x.

While a Lipschitz condition is sufficient to obtain uniqueness and must be imposed if we are to study differential equations with parameters, certain results, in particular Lemmas 1, 2 and 3 below, can be established without its use. These results will then be somewhat more general than required for our immediate purposes.

2.2 In the last theorem we obtained the existence of the Picard iterants $y^{(k)}(x)$ in the case of one unknown function. We are now concerned with the case of many dependent variables, that is, with systems of differential equations of the form of Equation (9). We write, from Equation (9),

$$y_i^{(k+1)}(x) = y_{i,0} + \int_{x_0}^{x} f_i(y_1^{(k)}(t), \ldots, y_n^{(k)}(t), t)\, dt, \; i = 1, \ldots, n. \quad (10)$$

When all terms of the above equation have been suitably defined, the $y_i^{(k+1)}(x)$ functions are known as the *Picard iterants* in n dependent variables.

Our first task is to show that under the usual assumptions regarding the f_i, the $y_i^{(k+1)}$ are well defined. That is, we shall show that $(y_1^{(k)}(x), \ldots, y_n^{(k)}(x), x)$ is in the domain of definition of the f_i. The lemma we shall prove (Lemma 1) is analogous (but slightly more general) than the construction employed in Theorem 1.

Lemma 1. Hypothesis. (i) Let $f_i(y_1, \ldots, y_n, x)$ be n real valued functions of the $n + 1$ real variables $y_1, \ldots, y_n$, x defined and continuous on a convex open region $\mathfrak{A}$ of $(n + 1)$-dimensional euclidean space.

(ii) Let $(y_{1,0}, \ldots, y_{n,0}, x_0)$ be a point in $\mathfrak{A}$ and $\mathfrak{R}$ a neighborhood of this point defined by the inequalities $|y_j - y_{j,0}| \leq a$, $|x - x_0| \leq a$, $\mathfrak{R} \subset \mathfrak{A}$.

(iii) Let M be a constant such that for every $(y_1, \ldots, y_n, x) \in \mathfrak{R}$, $|f_i(y_1, \ldots, y_n, x)| \leq M$, $i = 1, \ldots, n$.

(iv) Let $(y_1', \ldots, y_n', x')$ be any interior point of $\mathfrak{R}$. Let b_1 be the distance along the x-axes from this point to the boundary of $\mathfrak{R}$, that is

$$b_1 = \min\,[(x_0 + a - x'), (x' - x_0 + a)].$$

Let

$$c_1 = \min_j\,[(y_{j,0} + a - y_j'), (y_j' - y_{j,0} + a)].$$

Let

$$b_2 = \min (b_1, c_1/M).$$

(v) Let $y_1(x), \ldots, y_n(x)$ be n continuous functions of x known to be defined for $|x - x'| \leq b_2$ and such that for $|x - x'| \leq b_2$, $(y_1(x), \ldots, y_n(x), x) \varepsilon \mathfrak{R}$ and such that $y_j(x') = y_j'$, $j = 1, \ldots, n$.

(vi) Let

$$z_j(x) = y_j' + \int_{x'}^{x} f_j(y_1(t), \ldots, y_n(t), t)\, dt, \qquad j = 1, \ldots, n.$$

Conclusion. The $z_j(x)$ functions are defined and continuous for $|x - x'| \leq b_2$ and $(z_1(x), \ldots, z_n(x), x) \varepsilon \mathfrak{R}$ for $|x - x'| \leq b_2$.

Proof. By (v) of the hypothesis, for

$$|x - x'| \leq b_2, \qquad (y_1(x), \ldots, y_n(x), x) \varepsilon \mathfrak{R}$$

and hence in $\mathfrak{A}$. Also by (v) the $y_j(x)$ are continuous for $|x - x'| \leq b_2$. By (i) of the hypothesis the f_i are continuous on $\mathfrak{A}$ and hence the $f_i(y_1(x), \ldots, y_n(x), x)$, $i = 1, \ldots, n$ are continuous functions of x for $|x - x'| \leq b_2$. Hence, f_i, considered as a function of x, is Riemann integrable and we can define the (necessarily continuous) functions

$$z_j(x) = y_j' + \int_{x'}^{x} f_j(y_1(t), \ldots, y_n(t), t)\, dt, \qquad j = 1, \ldots, n,$$

for $|x - x'| \leq b_2$.

For $|x - x'| \leq b_2$ we have

$$\begin{aligned} |z_j(x) - y_j'| &= \left| \int_{x'}^{x} f_j(y_1(t), \ldots, y_n(t), t)\, dt \right| \leq M\,|x - x'| \\ &\leq Mb_2 \leq c_1 \leq \min_j \; [(y_{j,0} + a - y_j'), (y_j' - y_{j,0} + a)] \qquad (11) \end{aligned}$$

From Equation (11)

$$\begin{aligned} z_j(x) - y_j' &\leq y_{j,0} + a - y_j' \\ z_j(x) &\leq y_{j,0} + a. \qquad (12) \end{aligned}$$

Also from Equation (11)

$$\begin{aligned} y_j' - z_j(x) &\leq y_j' - y_{j,0} + a \\ z_j(x) &\geq y_{j,0} - a. \qquad (13) \end{aligned}$$

Equations (12) and (13) imply

$$|z_j(x) - y_{j,0}| \leq a, \qquad j = 1, \ldots, n.$$

Now

$$|x - x'| \leq b_2 \leq b_1 \leq \min [(x_0 + a - x'), (x' - x_0 + a)],$$

and by the same technique used above we conclude

$$|x - x_0| \leq a.$$

We see therefore that for $|x - x'| \leq b_2$, we have $|z_j(x) - y_{j,0}| \leq a$ and $|x - x_0| \leq a$, and hence the point $(z_1(x), \ldots, z_n(x), x)$ is in $\mathfrak{R}$ for $|x - x'| \leq b_2$.

In general, Lemma 1 can be used to show that the successive Picard iterants can be defined (cf. Lemma 2). As a corollary to Lemma 1, we note that if $x' = x_0$, $y_j' = y_{j,0}$, $b_1 = a$, $c_1 = a$, $b_2 = \min (a, a/M)$ then the lemma becomes the exact analog of the construction used in Theorem 1. However, the more general form of the above lemma will be convenient in future developments.

2.3 We now wish to prove the existence of the Picard iterants on a common x range for a specified initial point.

Lemma 2. Hypothesis. (i) Same as Lemma 1, (i)

(ii) Same as Lemma 1, (ii)

(iii) Same as Lemma 1, (iii)

(iv) Let

$$y_j^{(0)}(x) = y_{j,0}, \qquad j = 1, \ldots, n$$

and let $y_1^{(k+1)}(x), \ldots, y_n^{(k+1)}(x)$ (the Picard iterants) be defined recursively by the equations

$$y_i^{(k+1)}(x) = y_{i,0} + \int_{x_0}^{x} f_i(y_1^{(k)}(t), \ldots, y_n^{(k)}(t), t)\, dt, \qquad i = 1, \ldots, n.$$

(v) Let

$$b^* = \min \left(a, \frac{a}{M}\right).$$

Conclusion. All the $y_i^{(k+1)}(x)$, $i = 1, \ldots, n$, $k = 0, 1, 2, \ldots$ are defined and continuous for $|x - x_0| \leq b^*$.

Proof. Consider Lemma 1 with $x' = x_0$, $y_j' = y_{j,0}$, $y_j(x) = y_j^{(0)}(x)$, $j = 1, \ldots, n$. Then $b_1 = a$, $c_1 = a$, $b_2 = b^*$. Consequently $z_j(x) = y_j^{(1)}(x)$ and the $y_1^{(1)}(x), \ldots, y_n^{(1)}(x)$ are defined and continuous for $|x - x_0| \leq b^*$. If we assume that the $y_1^{(k)}(x), \ldots, y_n^{(k)}(x)$ are defined and continuous for $|x - x_0| \leq b^*$, then we can prove inductively that the $y_1^{(k+1)}(x), \ldots, y_n^{(k+1)}(x)$ are also defined and continuous for $|x - x_0| \leq b^*$. For, let $x' = x_0$, $y_j' = y_{j,0}$, $y_j(x) = y_j^{(k)}(x)$. Then $b_1 = a$, $c_1 = a$, $b_2 = b^*$ and hence $z_j(x) = y_j^{(k+1)}(x)$, which proves the lemma.

2.4 As has already been pointed out, the above lemma shows the existence of the Picard iterants on a common x range for a *specified* initial point. On the other hand, it is highly desirable to have a common range for the Picard iterants for a *range* of initial conditions as well. This we shall do in Lemma 3.

Lemma 3. Hypothesis. (i) Same as Lemma 2 (i) — (v) inclusive.

(ii) Let $b^+ = \frac{1}{3} b^* = \frac{1}{3} \min \left(a, \frac{a}{M}\right)$

(iii) Let $(y_1', \ldots, y_n', x')$ be a point in $\mathfrak{R}$ such that $|y_j' - y_{j,0}| \leq \frac{1}{3} a$ and $|x' - x| \leq b^+$.

Conclusion. If we define the Picard iterants as in (iv) of Lemma 2 with the initial iterant $y_j'^{(0)}(x) = y_j'$, then the $y_j'^{(k)}(x)$ are defined for $|x - x_0| \leq b^+$ for each such initial value $y_1', \ldots, y_n', x'$.

Proof. Consider Lemma 1 and let the $(y_1', \ldots, y_n', x')$ of the present lemma coincide with the $(y_1', \ldots, y_n', x')$ of Lemma 1. Since $|x' - x_0| \leq b^+ \leq \frac{a}{3}$ by hypothesis, any b_1 for Lemma 1 which is greater than or equal to $\frac{2a}{3}$ is adequate. Similarly, since $|y_j' - y_{j,0}| \leq \frac{a}{3}$ we may take

$$c_1 \geq \frac{2}{3} a.$$

Therefore,

$$b_2 \geq \min\left(\frac{2}{3} a, \frac{2}{3}\frac{a}{M}\right) = \frac{2}{3} \min\left(a, \frac{a}{M}\right) = 2\, b^+.$$

Now Lemma 1 shows that the first iterant is defined for $|x - x'| \leq b_2$ for each $(y_1', \ldots, y_n', x')$. And by induction (as in Lemma 2) we conclude all the iterants are defined for $|x - x'| \leq b_2$. But the two conditions $|x - x_0| \leq b^+$ and $|x' - x_0| \leq b^+$ imply $|x - x'| \leq 2b^+ \leq b_2$.

3. The Picard existence theorem

3.1 So far, we have used only the continuity of the f_i functions on $\mathfrak{A}$ to establish the existence of the Picard iterants,

$$y_j^{(k)}(x) = y_{j,0} + \int_{x_0}^{x} f_i(y_1^{(k-1)}(t), \ldots, y_n^{(k-1)}(t), t)\, dt$$

over the interval $|x - x_0| \leq b^*$.

The Lipschitz condition, as was discussed above, will be used to establish the uniform convergence of the Picard iterants on $|x - x_0| \leq b^*$, and the fact that the limit functions constitute a solution of the given differential system

$$\frac{dy_i}{dx} = f_i(y_1, \ldots, y_n, x), \qquad i = 1, \ldots, n$$

through the initial point $(y_{1,0}, \ldots, y_{n,0}, x_0)$ over the interval $|x - x_0| \leq b^*$.

3.2 The Picard theorem, which establishes existence in the small is stated in Theorem 2 below. It is a less general existence theorem than the Peano theorem since we are assuming a Lipschitz condition. However, as frequently remarked above we have other reasons for introducing the Picard iterants than just to prove an existence theorem in the small.

Theorem 2. Hypothesis. (i) Let $f_i(y_1, \ldots, y_n, x)$ be n real valued functions of the $n + 1$ real variables $y_1, \ldots, y_n, x$ defined and continuous on a convex open region $\mathfrak{A}$ of $(n + 1)$-dimensional euclidean space.

(ii) (Lipschitz condition.) There exists a constant L such that if $(y_1^*, \ldots, y_n^*, x)$ and $(y_1^+, \ldots, y_n^+, x)$ are two points of $\mathfrak{A}$, then

$$|f_i(y_1^*, \ldots, y_n^*, x) - f_i(y_1^+, \ldots, y_n^+, x)| \leq L \sum_{j=1}^{n} |y_j^* - y_j^+|, \qquad i = 1, \ldots, n.$$

(iii) Let $(y_{1,0}, \ldots, y_{n,0}, x_0)$ be a point of $\mathfrak{A}$.

Conclusion. The Picard iterants

$$y_i^{(0)}(x) = y_{1,0}$$

$$y_i^{(k+1)}(x) = y_{1,0} + \int_{x_0}^{x} f_i(y_1^{(k)}(t), \ldots, y_n^{(k)}(t), t)\, dt, \qquad \begin{array}{l} i = 1, \ldots, n \\ k = 0, 1, 2, \ldots \end{array}$$

(whose existence for a common x range $|x - x_0| \leq b^*$ was shown in Lemma 2) converge uniformly to continuous functions $y_i(x)$ which have the following properties:

(i) $y_{i,0} = y_i(x_0)$

(ii) $\frac{dy_i}{dx} = f_i(y_1(x), \ldots, y_n(x), x), \quad i = 1, \ldots, n$ for $|x - x_0| \leq b^*$.

(iii) The solutions $y_1(x), \ldots, y_n(x)$ are unique.

3.3 We first prove a lemma that will give us an upper bound on the Picard iterants.

Since $\mathfrak{A}$ is an open region, we can determine a rectangular neighborhood $\mathfrak{R}$, $|x - x_0| \leq a$, $|y_j - y_{j,0}| \leq a$, of the initial point $(y_{1,0}, \ldots, y_{n,0}, x_0)$ on which

$$|f_i(y_1, \ldots, y_n\, x)| \leq M, \qquad i = 1, \ldots, n,$$

and Lemma 2 shows that we may take

$$b^* = \min\left(a, \frac{a}{M}\right).$$

Let $\mathfrak{N}$ denote the x neighborhood $|x - x_0| \leq b^*$. The first iterant may be written

$$y_i^{(1)}(x) = y_{i,0} + \int_{x_0}^{x} f_i(y_1^{(0)}(t), \ldots, y_n^{(0)}(t), t)\, dt, \qquad i = 1, \ldots, n$$

for any $x \varepsilon \mathfrak{N}$. Since $y_i^{(0)}(t) = y_{i,0}$, this may be written in the form

$$y_i^{(1)}(x) - y_{i,0} = \int_{x_0}^{x} f_i(y_{1,0}, \ldots, y_{n,0}, t)\, dt$$

and hence

$$|y_i^{(1)}(x) - y_{i,0}| \leq \left| \int_{x_0}^{x} |f_i(y_{1,0}, \ldots\ y_{n,0}, t)|\, dt \right|$$
$$\leq M\, |x - x_0| \leq Mb^*, \qquad i = 1, \ldots, n.$$

We shall now show inductively that

$$|y_i^{(k)}(x) - y_i^{(k-1)}(x)| \leq M(Ln)^{k-1} \frac{(x - x_0)^k}{k!}, \qquad i = 1, \ldots, n.$$

For $k = 1$, this is the result proved above. Now assume the inequality for k. Then

$$|y_i^{(k+1)}(x) - y_i^{(k)}(x)|$$
$$\leq \left| \int_{x_0}^{x} |f_i(y_1^{(k)}, \ldots, y_n^{(k)}, t) - f_i(y_1^{(k-1)}, \ldots, y_n^{(k-1)}, t)|\, dt \right|$$

and by the Lipschitz condition

$$|y_i^{(k+1)} - y_i^{(k)}| \leq \left| \int_{x_0}^{x} L \sum_{j=1}^{n} |y_j^{(k)} - y_j^{(k-1)}|\, dt \right|.$$

Assume $x \geqq x_0$. (A similar proof holds for $x < x_0$.) Then

$$| y_i^{(k+1)} - y_i^{(k)} | \leqq Ln \int_{x_0}^{x} M(Ln)^{k-1} \frac{(t-x_0)^k}{k!} dt = M(Ln)^k \frac{(x-x_0)^{k+1}}{(k+1)!}.$$

This is the inequality for $k + 1$ and thus our induction is complete. Since $| x - x_0 | \leqq b^*$ we have

$$| y_i^{(k+1)} - y_i^{(k)} | \leqq M \frac{(Ln)^k (b^*)^{k+1}}{(k+1)!}, \qquad i = 1, \ldots, n.$$

Thus the following lemma has been proved:

Lemma 4. Under the hypothesis of Theorem 2,

$$| y_i^{(k)}(x) - y_i^{(k-1)}(x) | \leqq \frac{M (Lnb^*)^k}{(Ln)k!} \tag{14}$$

for any $x \varepsilon \mathfrak{R}$ and any i, $i = 1, \ldots, n$.

We note that since the right hand side of the inequality of Equation (14) is independent of x, the $| y_i^{(k)} - y_i^{(k-1)} |$ converge uniformly to zero on the interval $\mathfrak{R}$.

3.4 We wish to apply the Cauchy convergence criterion to show that the $y_i^{(k)}(x)$ converge uniformly. (It could also be done by the methods used in the proof of Theorem 1.) Hence we must show that $| y_j^{(p)} - y_j^{(q)} | \to 0$ for p, q sufficiently large. Lemma 4 establishes this fact only for $p = q + 1$.

Lemma 5. Hypothesis as in Theorem 2.

Conclusion. Let $p > q$. Then for every $j, j = 1, \ldots, n$ and each $x \varepsilon \mathfrak{R}$.

$$| y_j^{(p)}(x) - y_j^{(q)}(x) | \leqq \frac{M}{Ln} \mathcal{R}_{q+1} [e^{Lnb^*}]$$

where $\mathcal{R}_{q+1}[e^{Lnb^*}]$ is the remainder after $q + 1$ terms of the power series expansion for e^{Lnb^*}.

Proof. Since $p > q$ by hypothesis,

$$| y_j^{(p)} - y_j^{(q)} | = | y_j^{(p)} - y_j^{(p-1)} + y_j^{(p-1)} - \ldots + y_j^{(q+1)} - y_j^{(q)} |$$
$$\leqq \sum_{k=q+1}^{p} | y_j^{(k)} - y_j^{(k-1)} | \leqq \frac{M}{Ln} \sum_{k=q+1}^{p} \frac{(Lnb^*)^k}{k!} \leqq \frac{M}{Ln} \sum_{k=q+1}^{\infty} \frac{(Lnb^*)^k}{k!}.$$

3.5 From the fact that $\mathcal{R}_q \to 0$ as $q \to \infty$ independent of x we infer from the Cauchy convergence theorem that,

Lemma 6. The $y_j^{(p)}(x)$ functions converge uniformly on $\mathfrak{R}$ as $p \to \infty$ for each j.

Since each sequence $\{y_j^{(p)}(x)\}$ is a sequence of uniformly convergent *continuous* functions, we know that the limit function $y_j(x)$ is also continuous.

3.6 The final step in our proof of Theorem 2 is to show that the $y_j(x)$ are solutions of the given differential system on $\mathfrak{R}$ satisfying the initial conditions. This follows immediately from Lemma 7.

Lemma 7. Let $y_i^{(p)}(x)$ be the sequence of uniformly convergent functions on $\mathfrak{R}$ which have the limit $y_i(x)$, $i = 1, \ldots, n$. Then

$$y_i(x) = y_{i,0} + \int_{x_0}^{x} f_i(y_1(t), \ldots, y_n(t), t)\, dt, \qquad i = 1, \ldots, n.$$

Proof. We have, by definition of the Picard iterants

$$y_i^{(k+1)}(x) = y_{i,0} + \int_{x_0}^{x} f_i(y_1^{(k)}(t), \ldots, y_n^{(k)}(t), t)\, dt. \tag{15}$$

First we assert that

$$\lim_{k \to \infty} f_i(y_1^{(k)}(x), \ldots, y_n^{(k)}(x), x) = f_i(y_1(x), \ldots, y_n(x), x)$$

uniformly for any $x \in \mathfrak{R}$ and any i. To prove this, let $\varepsilon > 0$ be assigned. Then since $f_i(y_1, \ldots, y_n, x)$ is uniformly continuous for $(y_1, \ldots, y_n, x) \in \mathfrak{R}$ (cf. Lemma 1), there exists a uniform δ such that for all i,

$$| f_i(y_1^{(k)}(x), \ldots, y_n^{(k)}(x), x) - f_i(y_1(x), \ldots, y_n(x), x) | < \varepsilon$$

whenever $| y_j^{(k)}(x) - y_j(x) | < \delta$ and (trivially) $| x - x | < \delta$. Now since the $y_j^{(k)}(x)$ converge uniformly, we can, with the above δ, find a K, such that if $k > K$,

$$| y_j^{(k)}(x) - y_j(x) | < \delta, \qquad j = 1, \ldots, n,$$

for all $x \in \mathfrak{R}$. Hence, given an $\varepsilon > 0$, there exists a K, such that if $k > K$,

$$| f_i(y_1^{(k)}(x), \ldots, y_n^{(k)}(x), x) - f_i(y_1(x), \ldots, y_n(x), x) | < \varepsilon$$

for all i, $i = 1, \ldots, n$ and all $x \in \mathfrak{R}$ — which proves our assertion.

Now

$$\lim_{k \to \infty} \int_{x_0}^{x} f_i(y_1^{(k)}(t), \ldots, y_n^{(k)}(t), t)\, dt = \int_{x_0}^{x} f_i(y_1(t), \ldots, y_n(t), t)\, dt$$

since the integral of the limit of a sequence of uniformly convergent functions is equal to the limit of the integral. Using this result and taking the limit of both sides of Equation (15) we obtain the desired result.

3.7 Theorem 2 is now proved. For, from Lemma 10 of Chapter 1 the integral equations

$$y_i(x) = y_{i,0} + \int_{x_0}^{x} f_i(y_1(t), \ldots, y_n(t), t)\, dt, \qquad i = 1, \ldots, n$$

are equivalent to the differential equations,

$$\frac{dy_i}{dx} = f_i(y_1(x), \ldots, y_n(x), x), \qquad i = 1, \ldots, n. \tag{16}$$

Clearly, $y_i(x_0) = y_{i,0}$. Finally, the solution is unique on $\mathfrak{R}$ by the Lipschitz condition. (Cf. Theorem 2 of Chapter 3.)

4. Continuity in initial conditions

4.1 As we have already seen, the solution of a given system of differential equations varies with the initial conditions. Now if we have a Lipschitz condition on a region $\mathfrak{A}$, then to each point of $\mathfrak{A}$ there is a *unique* solution, valid in an interval, which goes through this point. We thus have a one-to-one correspondence between initial points and solutions which we write as

$$y_i(x) = y_i(x, y_1^*, \ldots, y_n^*, x^*) \tag{17}$$

where the solution $y_j(x)$ of the system of differential equations, Equation (16), is a function of x and of the parameters (representing initial conditions) $y_1^*, \ldots, y_n^*, x^*$. We emphasize again the fact that to study the dependence of solutions on initial conditions we must have uniqueness, for without uniqueness we would not have such a functional relationship as Equation (17).

Our next task is to prove (Theorem 3) that under certain conditions, and for a certain neighborhood $\mathfrak{N}$ of the point $y_{1,0}, \ldots, y_{n,0}, x_0$;

$$y_j(x) = y_j(x, y_1^*, \ldots, y_n^*, x^*)$$

is a continuous function of $x, y_1^*, \ldots, y_n^*, x^*$ for $(y_1^*, \ldots, y_n^*, x^*) \in \mathfrak{N}$. We shall use the theory of the Picard iterants to show that each iterant $y_j^{(k)}(x, y_1^*, \ldots, y_n^*, x^*)$ is jointly continuous in its $n + 2$ variables and that the convergence of the $y_j^{(k)}$ is uniform in all variables.

4.2 The statement of the theorem on the continuous dependence of solutions on initial condition parameters is:

Theorem 3. Hypothesis. (i) Let $f_i(y_1, \ldots, y_n, x)$ be n real valued functions of the $n + 1$ real variables $y_1, \ldots, y_n, x$ defined and continuous on a convex open region $\mathfrak{A}$ of $(n + 1)$-dimensional euclidean space.

(ii) Let the $f_i(y_1, \ldots, y_n, x)$ functions satisfy a Lipschitz condition with constant L in $\mathfrak{A}$.

(iii) Let $(y_{1,0}, \ldots, y_{n,0}, x_0)$ be a point in $\mathfrak{A}$ and $\mathfrak{R}$ a neighborhood of this point defined by the inequalities $|y_j - y_{j,0}| \leq a$, $|x - x_0| \leq a$, with $\mathfrak{R} < \mathfrak{A}$.

(iv) Let M be a constant such that for every $(y_1, \ldots, y_n, x) \varepsilon \mathfrak{R}$, $|f_i(y_1, \ldots, y_n, x)| \leq M$, $i = 1, \ldots, n$.

(v) Let $b^+ = \frac{1}{3} \min\left(a, \frac{a}{M}\right)$.

Conclusion. There exist n functions

$$y_j = \varphi_j(x, y_1^*, \ldots, y_n^*, x^*)$$

defined on a region $\mathfrak{M}$, $|x - x_0| \leq b^+$, $|y_j^* - y_{j,0}| \leq b^+$, $|x^* - x_0| \leq b^+$, such that

(i) $\partial\varphi_j/\partial x = f_j(\varphi_1, \ldots, \varphi_n, x)$ for $|x - x_0| \leq b^+$, $j = 1, \ldots, n$.

(ii) The φ_j are jointly continuous in the $n + 2$ variables $x, y_1^*, \ldots, y_n^*, x^*$ on $\mathfrak{M}$, $j = 1, \ldots, n$.

(iii) $y_j^* = \varphi_j(x^*, y_1^*, \ldots, y_n^*, x^*)$.

(iv) The functions $\varphi_1, \ldots, \varphi_n$ are specified uniquely by the above properties.

4.3 The proof of the above theorem is facilitated by the following function theoretic lemma.

Lemma 8. Hypothesis. Let $\psi(\xi, \lambda_1, \ldots, \lambda_r)$ be a real valued function of the $r + 1$ real variables $\xi, \lambda_1, \ldots, \lambda_r$ defined and continuous on an open region $\mathfrak{S}$ of $(r + 1)$-dimensional euclidean space. Let $(\bar{\xi}, \bar{\lambda}_1, \ldots, \bar{\lambda}_r)$ be a point in $\mathfrak{S}$ and $\mathfrak{T}$ a rectangular neighborhood of this point contained in $\mathfrak{S}$, and defined by the inequalities $|\bar{\xi} - \xi| \leq \alpha$, $|\bar{\lambda}_i - \lambda_i| \leq \alpha$, $i = 1, \ldots, r$.

Conclusion.

$$\Psi(\xi, \lambda_1, \ldots, \lambda_r) = \int_{\bar{\xi}}^{\xi} \psi(\zeta, \lambda_1, \ldots, \lambda_r)\, d\zeta$$

is jointly continuous in $(\xi, \lambda_1, \ldots, \lambda_r)$ for $(\xi, \lambda_1, \ldots, \lambda_r) \varepsilon \mathfrak{T}$.

Proof. We must show that given an $\varepsilon > 0$ there exists a $\delta > 0$ such that

$$|\Psi(\xi', \lambda_1', \ldots, \lambda_r') - \Psi(\xi'', \lambda_1'', \ldots, \lambda_r'')| \leqq \varepsilon$$

whenever $|\xi' - \xi''| < \delta$, $|\lambda_i' - \lambda_i''| < \delta$, $i = 1, \ldots, r$. Let ε be assigned. Since ψ is continuous on $\mathfrak{T}$ it is bounded, say with bound M, and uniformly continuous. Hence, given an ε_0 there exists a δ_0 such that

$$|\psi(\xi', \lambda_1', \ldots, \lambda_r') - \psi(\xi'', \lambda_1'', \ldots, \lambda_r'')| < \varepsilon_0$$

whenever $|\xi' - \xi''| < \delta_0$, $|\lambda_i' - \lambda_i''| < \delta_0$, $i = 1, \ldots, r$.

Let $\varepsilon_0 = \varepsilon/2a$ and let

$$\delta = \min\left(\delta_0, \frac{\varepsilon}{2M}\right).$$

Consider

$$\begin{aligned}\Psi(\xi', \lambda_1', \ldots, \lambda_r') - \Psi(\xi'', \lambda_1'', \ldots, \lambda_r'') \qquad (18)\\ = \int_{\bar{\xi}}^{\xi'} [\psi(\zeta, \lambda_1', \ldots, \lambda_r') - \psi(\zeta, \lambda_1'', \ldots, \lambda_r'')]\, d\zeta - \int_{\xi'}^{\xi''} \psi(\zeta, \lambda_1'', \ldots, \lambda_r'')\, d\zeta.\end{aligned}$$

Now

$$|\psi(\xi, \lambda_1', \ldots \lambda_r') - \psi(\xi, \lambda_1'', \ldots, \lambda_r'')| < \varepsilon_0 = \frac{\varepsilon}{2a}$$

whenever $|\lambda_i' - \lambda_i''| < \delta \leqq \delta_0$, $i = 1, \ldots, r$ and (trivially) $|\xi - \xi| < \delta \leqq \delta_0$. Hence, from Equation (18),

$$\begin{aligned}&|\Psi(\xi', \lambda_1', \ldots, \lambda_r') - \Psi(\xi'', \lambda_1'', \ldots, \lambda_r'')|\\ &\qquad \leqq \frac{\varepsilon}{2a}|\bar{\xi} - \xi'| + M|\xi' - \xi''| \leqq \frac{\varepsilon}{2a}a + M\frac{\varepsilon}{2M} = \varepsilon\end{aligned}$$

whenever $|\xi' - \xi''| < \delta$, $|\lambda_i' - \lambda_i''| < \delta$, $i = 1, \ldots, r$.

4.4 We now prove Theorem 3. Let $(y_1^*, \ldots, y_n^*, x^*)$ be a point of $\mathfrak{A}$ such that $|y_j^* - y_{j,0}| \leqq b^+$. By Lemma 3, the iterants

$$y_j^{(0)}(x) = y_j^*$$

$$y_j^{(k+1)}(x, y_1^*, \ldots, y_n^*, x^*) = y_j^* + \int_{x^*}^{x} f_j(y_1^{(k)}, \ldots, y_n^{(k)}, t)\, dt, \qquad (19)$$
$$j = 1, \ldots, n$$

are defined on $|x - x_0| \leqq b^+$ for every k and for every set of initial values $y_1^*, \ldots, y_n^*, x^*$ with $|y_j^* - y_{j,0}| \leqq b^+$, $|x^* - x_0| \leqq b^+$.

We see immediately that $y_j^{(o)}(x) \equiv y_j^*$ is jointly continuous in $x, y_1^*, \ldots, y_n^*, x^*$ on $\mathfrak{M}$. Hence $y_j^{(1)}$ is jointly continuous (f_j is continuous and apply Lemma 8). To prove that $y_j^{(k+1)}$ is jointly continuous in $x, y_1^*, \ldots, y_n^*, x^*$ on $\mathfrak{M}$ for $k = 0, 1, 2, \ldots,$ and any j, $j = 1, \ldots, n$, we proceed by induction. Suppose

$$y_j^{(k)}(x, y_1^*, \ldots, y_n^*, x^*)$$

has this property. Then $f_j(y_1^{(k)}, \ldots, y_n^{(k)}, t)$ is jointly continuous in the $n + 2$ variables $x, y_1^*, \ldots, y_n^*, x^*$, and by Lemma 8, so is $\int_{x^*}^{x} f_j \, dt$.

4.5 We next show that the $y_j^{(k)}(x, y_1^*, \ldots, y_n^*, x^*)$ functions converge uniformly on $\mathfrak{M}$ in the $n + 2$ arguments. From Lemma 4,

$$| y_i^{(k)}(x, y_1^*, \ldots, y_n^*, x^*) - y_i^{(k-1)}(x, y_1^*, \ldots, y_n^*, x^*) | \leq \frac{M(Lnb^+)^k}{(Ln)k!}$$

(using b^+ in place of b^*). The constants M, L, n, b^+ are independent of x. If we can show that they are also independent of y_j^*, x^* it will follow, exactly as in the proof of Lemmas 4, 5, 6, that the Picard iterants converge uniformly in all the variables together, and hence converge to continuous functions. Now $M \geq | f_i |$, L is the Lipschitz constant, n is the number of equations, $b^+ = \frac{1}{3} \min (a, a/M)$ and clearly these are independent of the y_j^* and x^*. What remains to be shown is done exactly as in the proof of the Picard theorem. (Cf. Lemma 7, Theorem 2.)

4.6 We now state the implicit case counterpart of Theorem 3. (Cf. Theorem 3 of Chapter 3.)

Theorem 4. Hypothesis. (i) Let

$$F_i(y_1', \ldots, y_n' \; y_1, \ldots, y_n, x) = 0, \qquad i = 1, \ldots, n$$

be n real valued functions of the $2n + 1$ real variables $y_1', \ldots, y_n',$ $y_1, \ldots, y_n, x$ defined and continuous on a convex open region $\mathfrak{A}$ of $(2n + 1)$-dimensional euclidean space.

(ii) $\dfrac{\partial F_i}{\partial y_j'}$ exist and are continuous on $\mathfrak{A}$ for $i, j = 1, \ldots, n$.

(iii) $\dfrac{\partial F_i}{\partial y_j}$ exist and are continuous on $\mathfrak{A}$ for $i, j = 1, \ldots, n$.

(iv) There exists a point $(y'_{1,0}, \ldots, y'_{n,0}, y_{1,0}, \ldots, y_{n,0}, x_0)$ in $\mathfrak{A}$ such that

$$F_i(y'_{1,0}, \ldots, y'_{n,0}, y_{1,0}, \ldots, y_{n,0}, x_0) = 0, \qquad i = 1, \ldots, n$$

and the Jacobian

$$J = \frac{\partial(F_1, \ldots, F_n)}{\partial(y'_1, \ldots, y'_n)} \neq 0$$

at this point.

Conclusion. There exist n functions

$$y_j = \varphi_j(x, y_1^*, \ldots, y_n^*, x^*)$$

defined on a region $\mathfrak{M}$ of the $n + 2$ variables $x, y_1^*, \ldots, y_n^*, x^*$ such that $\partial\varphi_j/\partial x$ is continuous and

$$\text{(i) } F_i\left(\frac{\partial\varphi_1}{\partial x}, \ldots, \frac{\partial\varphi_n}{\partial x}, \varphi_1, \ldots, \varphi_n, x\right) = 0, \quad i = 1, \ldots n$$

for $(x, y_1^*, \ldots, y_n^*, x^*) \varepsilon \mathfrak{M}$,

(ii) The φ_j are jointly continuous in the $n + 2$ variables $x, y_1^*, \ldots, y_n^*, x^*$ on $\mathfrak{M}$, $j = 1, \ldots, n$,

(iii) The φ_j are unique.

Since no new ideas are involved, the proof of this theorem will be left as an exercise for the reader. Note that, as in Theorem 3 of Chapter 3, the existence and continuity of the $\partial F_i/\partial y_j$ imply a Lipschitz condition.

4.7 Theorem 3 on the continuous dependence of solutions on initial conditions was a theorem "in the small." That is, we exhibited a b^+ neighborhood such that there existed solutions with initial conditions in this neighborhood which were continuous in these initial conditions. We would like a theorem in the large for solutions which are jointly continuous in x and the initial conditions. Examples of such theorems are furnished by the corollaries to Theorems 1 and 2 of Chapter 3 where part of the *hypothesis* included the existence of a solution on a (not necessarily small) x range, $x_1 \leq x \leq x_2$.

In general it is not always possible to push a solution defined "in the small" to a solution defined "in the large." (Note that in the above mentioned corollaries one solution in the large was always postulated.)

As an example, consider the differential equation

$$y' = y^2.$$

Its solution through the point (0,1) is

$$y = \frac{1}{1-x}.$$

We see that this solution through (0,1) is certainly defined for $0 \leq x < 1$, but this solution cannot be extended to $x = 1$. Hence, this solution has a singularity at $x = 1$, although the given differential equation gave us no hint as to its existence.

A theorem in the large can be proved for solutions that are continuous functions of x and the initial conditions together, *only if we are given that through some point one solution in the large exists.*

4.8 A theorem in the large on the continuous dependence of solutions on initial conditions parameters is given by Theorem 5, whose proof appears in Section 4.10.

Theorem 5. Hypothesis. (i) Let $f_i(y_1, \ldots, y_n, x)$ be n real valued functions of the $n + 1$ real variables $y_1, \ldots, y_n, x$ defined and continuous on a convex open region $\mathfrak{A}$ of $(n + 1)$-dimensional euclidean space.

(ii) The f_i functions satisfy a Lipschitz condition with constant L on $\mathfrak{A}$.

(iii) Let $(y_{1,0}, \ldots, y_{n,0}, x_0)$ be a point in $\mathfrak{A}$.

(iv) There exists a solution $y_j = \varphi_j(x)$, $j = 1, \ldots, n$ of the differential equations

$$\frac{dy_i}{dx} = f_i(y_1, \ldots, y_n, x)$$

defined on $x_0 \leq x \leq c$ with values $\{\varphi_1(x), \ldots, \varphi_n(x), x\}$ in $\mathfrak{A}$ and $y_{j,0} = \varphi_j(x_0)$.

Conclusion. There is a $b^+ > 0$ such that if $|y_j^* - y_{j,0}| \leq b^+$, $|x^* - x_0| \leq b^+$ there exist n functions

$$y_j = \psi_j(x, y_1^*, \ldots, y_n^*, x^*)$$

defined on $x_0 \leq x \leq c$ such that

(i) $\dfrac{\partial \psi_j}{\partial x} = f_j(\psi_1, \ldots, \psi_n, x)$ for $x_0 \leq x \leq c$,

(ii) $y_j^* = \psi_j(x^*, y_1^*, \ldots, y_n^*, x^*)$,

(iii) $\varphi_j(x) = \psi_j(x, y_{1,0}, \ldots, y_{n,0}, x_0)$,

(iv) The $\psi_j(x, y_1^*, \ldots, y_n^*, x^*)$ are jointly continuous in the $n + 2$ variables $(x, y_1^*, \ldots, y_n^*, x^*)$.

4.9 Note that (iv) of the hypothesis *postulates* the existence of a solution for $x_0 \leq x \leq c$. The number c is not necessarily to be thought of as small, cf. Figure 1 (which is drawn for the case $n = 1$).

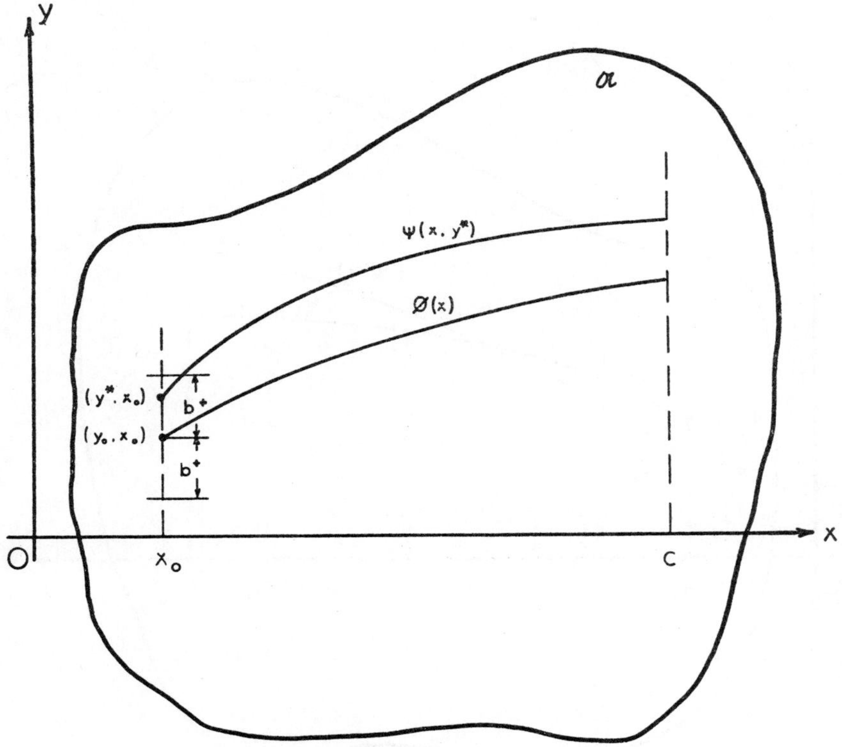

FIGURE 1

So we see that if we can get *one* solution running through $\mathfrak{A}$ with initial point $(x_0, y_{1,0}, \ldots, y_{n,0})$, then we can get infinitely many solutions running through $\mathfrak{A}$ with initial conditions $(x_0, y_1^*, \ldots, y_n^*, x^*)$ for $|y_j^* - y_{j,0}| \leq b^+$, $|x^* - x_0| \leq b^+$. In other words, given one solution in the large through a given point, we can find a region containing this solution which contains infinitely many solutions in the large.

As long as we have one solution in the large to tie things together, we can obtain other solutions. The real problem in the general non-linear cases dealt with here is to find this one solution. As the example given in Section 4.7 indicated, such a solution may not always exist in the large.

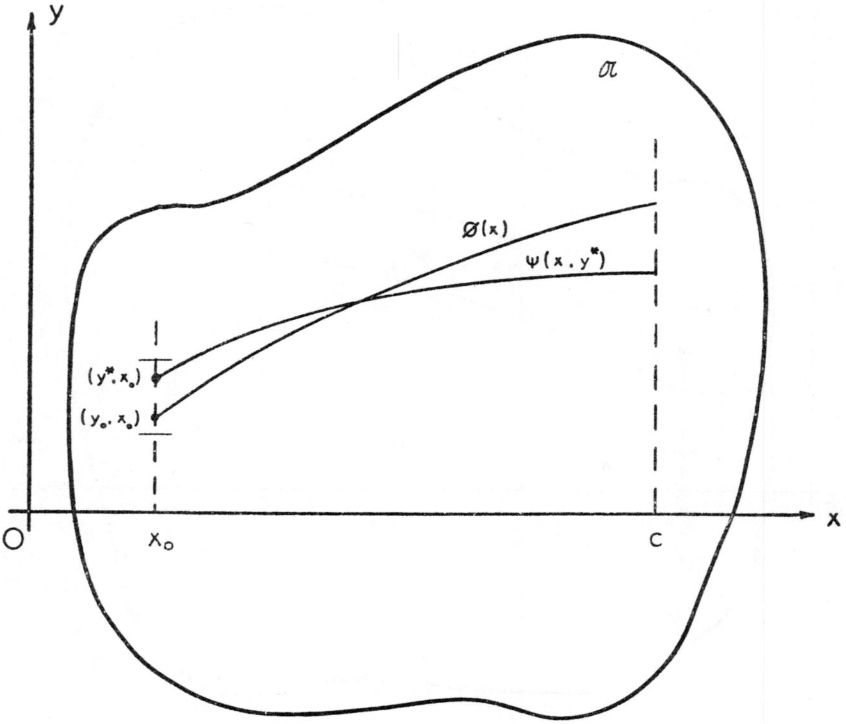

FIGURE 2

Returning to Figure 1, we see that uniqueness in the large means that two solutions cannot have a point in common. (Geometrically speaking, this means that Figure 2 is an impossible situation.) For the f_i satisfy a Lipschitz condition on $\mathfrak{A}$ and hence in the neighborhood of this common point, which yields a contradiction.

4.10 The proof of Theorem 5 depends on the Heine-Borel theorem. Consider the given solution $y_j = \varphi_j(x)$, $j = 1, \ldots, n$

which passes through $(y_{1,0}, \ldots, y_{n,0}, x_0)$. Since these functions are defined for $x_0 \leq x \leq c$ and $\mathfrak{A}$ is open we can find for every $\bar{x} \in [x_0, c]$ a set of y's, say $\bar{y}_1, \ldots, \bar{y}_n$ such that

$$(\bar{y}_1, \ldots, \bar{y}_n, x)$$

is in $\mathfrak{A}$. In fact, $(\varphi_1(\bar{x}), \ldots, \varphi_n(\bar{x}), \bar{x})$ is such a point. Since the hypothesis of Theorem 3 is satisfied for every $\bar{x} \in [x_0, c]$, there exists a neighborhood of each point, say $|\bar{x} - x| \leq b^+(\bar{x})$, ($b^+$ is a function of $\bar{x}$) such that there exist n functions

$$y_j = \psi_j(x, y_1^*, \ldots, y_n^*, x^*)$$

defined on $|\bar{x} - x| \leq b^+(\bar{x})$, $|\bar{y}_j - y_j^*| \leq b^+(\bar{x})$, $|\bar{x} - x^*| \leq b^+(\bar{x})$ such that

$$\text{(i)}\ \frac{\partial \psi_j}{\partial x} \equiv f_j(\psi_1, \ldots, \psi_n, x) \quad \text{for} \quad |\bar{x} - x| \leq b^+(\bar{x}),$$

(ii) The ψ_j are jointly continuous in the $n + 2$ variables x, $y_1^*, \ldots, y_n^*, x^*$,

(iii) $y_j^* = \psi_j(x^*, y_1^*, \ldots, y_n^*, x^*)$.

Now by the Heine-Borel theorem there exists a finite number of overlapping intervals which cover the x range $[x_0, c]$, say the b^+'s associated with the points $x_0 = \bar{x}_1, \ldots, \bar{x}_k = c$. We shall let $b^+(\bar{x}_\alpha)$ be the b^+ associated with the $\bar{x}_\alpha$ point, $\alpha = 1, \ldots, k$. Now starting with $\bar{x}_1$ we can determine a $b_1^+ > 0$ [which may be smaller than $b^+(\bar{x}_1)$] such that every solution with initial point $(y_1^*, \ldots, y_n^*, x^*)$ and with $|y_{j,0} - y_j^*| \leq b_1^+$, $|x_0 - x^*| \leq b_1^+$ passes into the second box [defined by $|y_j - \varphi_j(\bar{x}_2)| \leq b^+$, $|x - \bar{x}_2| \leq b^+(\bar{x}_2)$] along with the original solution φ_j. We may continue any such solution by means of a solution defined in the second box which agrees with this solution at the lower end point of the second box. Since these two solutions have a common point, the Lipschitz condition implies that they must be identical on the intersection of the first and second boxes. Of these solutions defined by b_1^+ we can find a set, determined in precisely the same way, by a $b_2^+ > 0$, [$b_2^+ \leq b_1^+$ and $b_2^+ \leq b^+(\bar{x}_2)$] such that the continuation of these solutions enters the third box along with the original solution φ_j of the hypothesis. As before, we can continue these solutions through the third box. We can continue this process until we have solutions defined through all k boxes with

$b_1^+ \geq b_2^+ \geq \ldots \geq b_k^+ = b^+ > 0$. This b^+ is the b^+ of the conclusion of Theorem 5.

It remains but to prove that the solutions

$$y_j = \psi_j(x, y_1^*, \ldots, y_n^*, x^*)$$

running from $x = x_0$ to $x = c$ are continuous in the initial condition parameters $y_1^*, \ldots, y_n^*, x^*$. By Theorem 3, the solutions ψ_j are continuous in $y_1^*, \ldots, y_n^*, x^*$ in the first $b^+(\bar{x}_1)$ box. Their values y_j^{**} at the beginning of the second box are taken as initial conditions for this box. Again by Theorem 3, the solution in the second box is continuous in the initial conditions, y_j^{**}, x^{**}. But a continuous function of a continuous function is continuous, and hence the solution in the second box is continuous in the original $y_1^*, \ldots, y_n^*, x^*$ parameters. Proceeding in this fashion we obtain the continuity of $y_j = \psi_j(x, y_1^*, \ldots, y_n^*, x^*)$ in the original $y_1^*, \ldots, y_n^*, x^*$ initial condition parameters throughout $[x_0, c]$.

5. Continuity in parameters

5.1 We have discussed in Theorems 3 and 5 the dependence of solutions on initial condition parameters. Both proofs involved the Picard iterant mechanism. These results can be generalized to another case of considerable interest, namely, the case in which parameters $\lambda_1, \ldots, \lambda_r$ appear in the f_i's themselves.

Up until the present we have considered systems of the form

$$\frac{dy_i}{dx} = f_i(y_1, \ldots, y_n, x).$$

Now we wish to consider systems of the form

$$\frac{dy_i}{dx} = f_i(y_1, \ldots, y_n, x, \lambda_1, \ldots, \lambda_r)$$

where the parameters $\lambda_1, \ldots, \lambda_r$ appear in the f_i.

We shall see that if we require the f_i to be continuous in the $n + r + 1$ variables $y_1, \ldots, y_n, x, \lambda_1, \ldots, \lambda_r$ in some region $\mathfrak{A}$ and satisfy a Lipschitz condition, then we can show that the solutions are continuous functions of the parameters. Precisely formulated:

Theorem 6. Hypothesis. (i) Let $f_i(y_1, \ldots, y_n, x, \lambda_1, \ldots, \lambda_r)$ be n real valued functions of the $n + 1 + r$ real variables $y_1, \ldots, y_n, x$,

$\lambda_1, \ldots, \lambda_r$ defined and continuous on a convex open region $\mathfrak{A}$ of $(n + 1 + r)$-dimensional euclidean space.

(ii) (Lipschitz condition.) There exists a constant L such that for every pair of points $(\bar{y}_1, \ldots, \bar{y}_n, x, \lambda_1, \ldots, \lambda_r)$ and $(y_1^+, \ldots, y_n^+, x, \lambda_1, \ldots, \lambda_r)$ in $\mathfrak{A}$ we have

$$| f_i(\bar{y}_1, \ldots, \bar{y}_n, x, \lambda_1, \ldots, \lambda_r) - f_i(y_1^+, \ldots, y_n^+, x, \lambda_1, \ldots, \lambda_r) | \leq L \sum_{j=1}^{n} | \bar{y}_j - y_j^+ |, \qquad i = 1, \ldots, n.$$

(iii) Let $(y_{1,0}, \ldots, y_{n,0}, x_0, \lambda_{1,0}, \ldots, \lambda_{r,0})$ be a point in $\mathfrak{A}$ and $\mathfrak{R}$, $\mathfrak{R} < \mathfrak{A}$, a neighborhood of this point defined by the inequalities

$$| y_j - y_{j,0} | \leq a, \quad | x - x_0 | \leq a, \quad | \lambda_i - \lambda_{i,0} | \leq a.$$

(iv) Let M be a constant such that for every

$$(y_1, \ldots, y_n, x, \lambda_1, \ldots, \lambda_r) \varepsilon \mathfrak{R},$$

$$| f_i(y_1, \ldots, y_n, x, \lambda_1, \ldots, \lambda_r) | \leq M, \quad i = 1, \ldots, n.$$

(v) Let $b^* = \min \left(a, \dfrac{a}{M}\right)$.

Conclusion. There exist n unique functions

$$y_j = \varphi_j(x, \lambda_1, \ldots, \lambda_r)$$

defined on a region $\mathfrak{M}$, $| x - x_0 | \leq b^*$, $| \lambda_i - \lambda_{i,0} | \leq a$ such that

(i) $\partial\varphi_i/\partial x = f_i(\varphi_1, \ldots, \varphi_n, x)$ for $| x - x_0 | \leq b^*$, $i = 1, \ldots, n$.

(ii) The φ_j are jointly continuous in the $r + 1$ variables $x, \lambda_1, \ldots, \lambda_r$ on $\mathfrak{M}$, $j = 1, \ldots, n$,

(iii) $y_{j,0} = \varphi_j(x_0, \lambda_1, \ldots, \lambda_r)$.

Proof. Define the Picard iterants by the equations

$$y_j^{(0)}(x) = y_{j,0}$$

$$y_j^{(k+1)}(x, \lambda_1, \ldots, \lambda_r) = y_{j,0} + \int_{x_0}^{x} f_j(y_1^{(k)}, \ldots, y_n^{(k)}, t, \lambda_1, \ldots, \lambda_r)\, dt,$$

$$j = 1, \ldots, n, \qquad k = 0, 1, 2, \ldots, \tag{20}$$

for $| x - x_0 | \leq b^*$, $| \lambda_i - \lambda_{i,0} | \leq a$. To show the Picard iterants are well defined we proceed as in Lemma 1. Clearly,

$$(y_1^{(0)}, \ldots, y_n^{(0)}, t, \lambda_1, \ldots, \lambda_r) \in \mathfrak{R}$$

for $|t - x_0| \leq b^*$, $|\lambda_i - \lambda_{i,0}| \leq a$. Hence

$$|f_j(y_1^{(0)}, \ldots, y_n^{(0)}, t, \lambda_1, \ldots, \lambda_r)| \leq M$$

and we deduce from Equation (20) with $k = 0$,

$$|y_j^{(1)}(x, \lambda_1, \ldots, \lambda_r) - y_{j,0}| \leq M|x - x_0| \leq Mb^* \leq M\left(\frac{a}{M}\right) = a.$$

Therefore, $(y_1^{(1)}, \ldots, y_n^{(1)}, x, \lambda_1, \ldots, \lambda_r) \in \mathfrak{R}$.

We can prove inductively that $(y_1^{(k)}, \ldots, y_n^{(k)}, x, \lambda_1, \ldots, \lambda_r) \in \mathfrak{R}$, $|x - x_0| \leq b^*$, $|\lambda_i - \lambda_{i,0}| \leq a$. For assume

$$(y_1^{(k-1)}, \ldots, y_n^{(k-1)}, x, \lambda_1, \ldots, \lambda_r) \in \mathfrak{R},$$

then $|f_i(y_1^{(k-1)}, \ldots, y_n^{(k-1)}, x, \lambda_1, \ldots, \lambda_r)| \leq M$, and from Equation (20)

$$y_j^{(k)}(x, \lambda_1, \ldots, \lambda_r) - y_{j,0} = \int_{x_0}^{x} f_i(y_1^{(k-1)}, \ldots, y_n^{(k-1)}, t, \lambda_1, \ldots, \lambda_r)\, dt.$$

And

$$|y_j^{(k)} - y_{j,0}| \leq M|x - x_0| \leq Mb^* \leq a.$$

We now have the existence of the Picard iterants on a common x range. By Lemma 8, $y_j^{(k+1)}(x, \lambda_1, \ldots, \lambda_r)$ [cf. Equation (20)] is jointly continuous in $(x, \lambda_1, \ldots, \lambda_r)$. Exactly as in Lemma 5, we show that for $p > q$

$$|y_i^{(p)}(x, \lambda_1, \ldots, \lambda_r) - y_i^{(q)}(x, \lambda_1, \ldots, \lambda_r)| \leq \frac{M}{Ln} \mathcal{R}_{q+1}[e^{Lnb^*}]. \quad (21)$$

Now M is the bound of the f_i on $\mathfrak{R}$ and therefore independent of $\lambda_1. \ldots, \lambda_r$. The Lipschitz constant L is also independent of the λ_i. By construction b^* is independent of the λ's. (Of course, M, L, n, b^* do not involve x.)

Hence, we deduce from Equation (21) that the $y_j^{(p)}(x, \lambda_1, \ldots, \lambda_r)$ converge uniformly in all $r + 1$ variables, and since $y_j^{(p)}(x, \lambda_1, \ldots, \lambda_r)$ is jointly continuous in the $r + 1$ variables, so is the limit function $y_j(x, \lambda_1, \ldots, \lambda_r)$. Lemma 7 completes the proof.

5.2 The above theorem can be generalized to include initial conditions as well as parameters. This is stated in Theorem 7 below.

Theorem 7. Hypothesis. (i) Let $f_i(y_1, \ldots, y_n, x, \lambda_1, \ldots, \lambda_r)$ be n real valued functions of the $n + 1 + r$ real variables $y_1, \ldots, y_n, x, \lambda_1, \ldots, \lambda_r$ defined and continuous on a convex open region $\mathfrak{A}$ of $(n + 1 + r)$-dimensional euclidean space.

(ii) (Lipschitz condition.) There exists a constant L such that for every pair of points

$$(\bar{y}_1, \ldots, \bar{y}_n, x, \lambda_1, \ldots, \lambda_r) \quad \text{and} \quad (y_1^+, \ldots, y_n^+, x, \lambda_1, \ldots, \lambda_r)$$

in $\mathfrak{A}$ we have

$$|f_i(\bar{y}_1, \ldots, \bar{y}_n, x, \lambda_1, \ldots, \lambda_r) - f_i(y_1^+, \ldots, y_n^+, x, \lambda_1, \ldots, \lambda_r)|$$
$$\leq L \sum_{j=1}^{n} |\bar{y}_j - y_j^+|, \qquad i = 1, \ldots, n.$$

(iii) Let $(y_{1,0}, \ldots, y_{n,0}, x_0, \lambda_{1,0}, \ldots, \lambda_{r,0})$ be a point in $\mathfrak{A}$.

Conclusion. There exist n functions

$$y_j = \varphi_j(x, y_1^*, \ldots, y_n^*, x^*, \lambda_1, \ldots, \lambda_r)$$

defined on a region $\mathfrak{M}$, $|x - x_0| \leq b^+$, $|y_j^* - y_{j,0}| \leq b^+$, $|x^* - x_0| \leq b^+$, $|\lambda_i - \lambda_{i,0}| \leq a$ such that

(i) $\partial\varphi_j/\partial x = f_j(\varphi_1, \ldots, \varphi_n, x)$ for $|x - x_0| \leq b^+$, $j = 1, \ldots, n$.

(ii) The φ_j are jointly continuous in the $n + r + 2$ variables $x, y_1^*, \ldots, y_n^*, x^*, \lambda_1, \ldots, \lambda_r$ on $\mathfrak{M}$, $j = 1, \ldots, n$.

(iii) $y_j^* = \varphi_j(x^*, y_1^*, \ldots, y_n^*, x^*, \lambda_1, \ldots, \lambda_r)$.

5.3 The analog of Theorem 5 for parameters, that is, existence in the large, is given in Theorem 8 below. As in the case of Theorem 6, we generalize it to include initial conditions as well.

Theorem 8. Hypothesis. (i) Let $f_i(y_1, \ldots, y_n, x, \lambda_1, \ldots, \lambda_r)$ be n real valued functions of the $n + 1 + r$ real variables $y_1, \ldots, y_n, x, \lambda_1, \ldots, \lambda_r$ defined and continuous on a convex open region $\mathfrak{A}$ of $(n + 1 + r)$-dimensional euclidean space.

(ii) The f_i functions satisfy a Lipschitz condition with constant L on $\mathfrak{A}$.

(iii) Let $(y_{1,0}, \ldots, y_{n,0}, x_0, \lambda_{1,0}, \ldots, \lambda_{r,0})$ be a point of $\mathfrak{A}$.

(iv) There exist solutions $\varphi_j(x)$, $j = 1, \ldots, n$ of the differential equations

$$\frac{dy_i}{dx} = f_i(y_1, \ldots, y_n, x, \lambda_{1,0}, \ldots, \lambda_{r,0}), \qquad i = 1, \ldots, n$$

which run through $\mathfrak{A}$, that is, are defined on $x_0 \leq x \leq c$ in $\mathfrak{A}$ and $y_{j,0} = \varphi_j(x_0)$.

Conclusion. There is a $b > 0$ such that if $|\lambda_i - \lambda_{i,0}| \leq b$, $|y_j^* - y_{j,0}| \leq b$ there exist n functions

$$y_j = \psi_j(x, y_1^*, \ldots, y_n^*, \lambda_1, \ldots, \lambda_r)$$

defined on $x_0 \leq x \leq c$ such that

(i) $\partial\psi_i/\partial x = f_i(\psi_1, \ldots, \psi_n, x, \lambda_1, \ldots, \lambda_r)$ for $x_0 \leq x \leq c, i = 1, \ldots, n$,

(ii) $y_i^* = \psi_i(x_0, y_1^*, \ldots, y_n^*, \lambda_1, \ldots, \lambda_r)$,

(iii) $\varphi_i(x) = \psi_i(x, y_{1,0}, \ldots, y_{n,0}, \lambda_{1,0}, \ldots, \lambda_{r,0})$,

(iv) The $\psi_i(x, y_1^*, \ldots, y_n^*, \lambda_1, \ldots, \lambda_r)$ are jointly continuous in the $n + r + 1$ variables $(x, y_1^*, \ldots, y_n^*, \lambda_1. \ldots, \lambda_r)$.

5.4 Analogous results are obtainable in the implicit case when a solution for a definite x-interval is known; or for the situation "in the small" under hypotheses which can be readily inferred from the above.

CHAPTER 5

Properties of Solutions

1. Introduction

In the previous chapter we considered the dependence of solutions on initial conditions and parameters. The functions $f_i(y_1, \ldots, y_n, x)$ or $f_i(y_1, \ldots, y_n, x, \lambda_1, \ldots, \lambda_r)$ were assumed continuous in some region $\mathfrak{A}$ and satisfied a Lipschitz condition in this region. The Lipschitz condition gave us uniqueness of solution which enabled us to obtain a functional relationship between solutions and initial conditions and parameters. We proved under these conditions that the solutions were jointly continuous in x and the parameters. It appears reasonable that if we assume stronger hypotheses regarding the f_i functions, additional results may be obtained.

In this direction we shall first assume that the partial derivatives $\partial f_i/\partial y_j$ exist and are continuous on $\mathfrak{A}$. As discussed in Section 1 of Chapter 3, this is a stronger condition on the f_i functions than a Lipschitz condition, that is, the existence and continuity of the partials implies a Lipschitz condition. On the other hand, a Lipschitz condition does not imply the existence of the partial derivatives. (Example: $y' = |y|$ in the neighborhood of $y = 0$. Clearly $|y|$ satisfies a Lipschitz condition while $\frac{\partial |y|}{\partial y}$ does not even exist at $y = 0$.) If we impose the partial derivative condition we can obtain additional results, for example, the fact that the solutions which depend on initial conditions can also be *differentiated* with respect to the y initial conditions.

As we proceed we shall make more and more restrictive assumptions on the f_i functions and thus in turn obtain the differentiability and even analyticity of solutions in initial conditions and parameters. Furthermore, the properties established here are not readily inferred from a numerical investigation. The range of definition of a solution, that is, its character in the large, can be indicated by a computation but a property such as "differentiability" relative to y_0 can ordinarily not be inferred.

2. Differentiation with respect to y initial condition for the case n = 1

2.1 To exhibit clearly the ideas involved we prove this theorem first for the simple case of one dependent variable.

Theorem 1. Hypothesis. (i) Let f(y, x) be a real valued function of the two real variables y and x defined and continuous on a convex open region $\mathfrak{A}$ of two dimensional euclidean space.

(ii) Let $\partial f/\partial y$ exist and be jointly continuous in y and x on $\mathfrak{A}$.

(iii) Let (y_0, x_0) be a point of $\mathfrak{A}$.

(iv) There exists a solution $\varphi(x)$ of the differential equation

$$\frac{dy}{dx} = f(y, x)$$

defined on $x_0 \leq x \leq c$ with values $\{\varphi(x), x\}$ in $\mathfrak{A}$ and with $y_0 = \varphi(x_0)$.

Conclusion. There exists a $b^+ > 0$ such that if $|y^* - y_0| \leq b^+$, $|x^* - x_0| \leq b^+$, then there exists a function

$$y = \varphi(x, y^*, x^*)$$

defined on $x_0 \leq x \leq c$ such that

(i) $\partial \varphi/\partial x \equiv f(\varphi, x)$ for $x_0 \leq x \leq c$, $|y^* - y_0| \leq b^+$, $|x^* - x_0| \leq b^+$.

(ii) $\varphi(x, y^*, x^*)$ is jointly continuous in x, y^*, x^*.

(iii) $y^* = \varphi(x^*, y^*, x^*)$.

(iv) $\partial \varphi/\partial y^*$ exists and is continuous for $x_0 \leq x \leq c$, $|y^* - y_0| < b^+$, $|x^* - x_0| \leq b^+$.

We note that the hypothesis of this theorem is identical with that of Theorem 5 of Chapter 4 except for (ii). In Theorem 5 of Chapter 4 we had a Lipschitz condition while here we have the existence and continuity of the partial derivative $\partial f/\partial y$. As discussed above, our present hypothesis is stronger, that is, implies the hypothesis of Theorem 5 of Chapter 4. Our conclusions (i)—(iii) are identical with those of Theorem 5 of Chapter 4 and, since our hypothesis is stronger, these results follow immediately. However, the crucial difference is (iv) of the conclusions which will follow because of the stronger (ii) of the hypothesis.

Note that in (iv) of the conclusions we defined an *open* region $|y^* - y_0| < b^+$ of the initial point y_0 instead of the closed region $|y^* - y_0| \leq b^+$ used in (i) — (iii). The reason for this is that we wish to consider Δy^* [in anticipation of forming the difference quotient $\frac{\Delta y}{\Delta y^*}\left(\rightarrow\frac{\partial y}{\partial y^*}\right)$] and we desire to have the new point $y^*+\Delta y^*$ remain in the region of definition of y^*.

The presence of parameters in f(y, x) would not affect the proof we are about to give and hence can be conveniently omitted for simplicity in notation.

2.2 We prove Theorem 1. Let (y^*, x^*) be a point such that $|y^*-y_0|<b^+$, $|x^*-x_0|\leq b^+$. (y^* is an *interior* point of the interval.) Consider an increment Δy^* taken so small that $y^*+\Delta y^*$ is in the b^+ neighborhood of y_0. We have then,

$$y(x)=\varphi(x, y^*, x^*) \quad \text{for} \quad x_0\leq x\leq c \tag{1}$$

and by definition

$$y(x)+\Delta y(x)=\varphi(x, y^*+\Delta y^*, x^*) \quad \text{for} \quad x_0\leq x\leq c. \tag{2}$$

These two functions, $y(x)$ and $y(x)+\Delta y(x)$, are solutions of the differential equation $y'=f(y, x)$ for $x_0\leq x\leq c$. Hence

$$\frac{dy}{dx}=f(y, x)$$

and

$$\frac{d}{dx}(y+\Delta y)=\frac{dy}{dx}+\frac{d}{dx}\Delta y=f(y+\Delta y, x)$$

and by subtraction

$$\frac{d}{dx}\Delta y=f(y+\Delta y, x)-f(y, x) \quad \text{for} \quad x_0\leq x\leq c.$$

Dividing by Δy^* (Δy^* cannot be zero since it is an increment) we have

$$\frac{d}{dx}\left(\frac{\Delta y}{\Delta y^*}\right)=\frac{f(y+\Delta y, x)-f(y, x)}{\Delta y^*}$$

since Δy^* is independent of x. We write

$$\begin{aligned}\frac{d}{dx}\left(\frac{\Delta y}{\Delta y^*}\right)&=\frac{f(y+\Delta y, x)-fy, x)}{\Delta y}\,\frac{\Delta y}{\Delta y^*} \quad &\text{if} \quad \Delta y\neq 0\\ &=0 \quad &\text{if} \quad \Delta y=0.\end{aligned} \tag{3}$$

For $\Delta y \neq 0$ we may write Equation (3) as

$$\frac{d}{dx}\left(\frac{\Delta y}{\Delta y^*}\right) = \frac{f\left(y + \frac{\Delta y}{\Delta y^*}\Delta y^*, x\right) - f(y, x)}{\frac{\Delta y}{\Delta y^*}\Delta y^*} \frac{\Delta y}{\Delta y^*}$$

and letting

$$z = \frac{\Delta y}{\Delta y^*}$$

we obtain

$$\begin{aligned} \frac{dz}{dx} &= \left[\frac{f(y + z\,\Delta y^*, x) - f(y, x)}{z\,\Delta y^*}\right] z, \quad z \neq 0 \\ &= 0 \qquad\qquad\qquad\qquad\qquad\qquad , \quad z = 0, \end{aligned} \tag{4}$$

which we shall write as

$$\frac{dz}{dx} = F(x, z, \Delta y^*), \qquad \Delta y^* \neq 0. \tag{5}$$

In Equation (5) y is a *known* function of x, that is $y = \varphi(x, y^*, x^*)$. Therefore, F is not to be regarded as a function of y. Equation (5) is to be regarded as a differential equation on the dependent variable z, with x as the independent variable and Δy^* a parameter. (F is also a function of y^* and x^* although we do not indicate these quantities explicitly.)

We *define* $F(x, z, \Delta y^*)$ for $\Delta y^* = 0$ as

$$F(x, z, 0) = f_y(y, x)\, z.^1 \tag{6}$$

Now $\partial f/\partial y$ is jointly continuous in y and x by hypothesis, and hence when y is replaced by the continuous function of x, $y = \varphi(x, y^*, x^*)$, $\partial f/\partial y$ becomes a continuous function of x. Clearly, $(\partial f/\partial y)\, z$ is jointly continuous in x and z.

Hence the differential equation on z,

$$\frac{dz}{dx} = F(x, z, \Delta y^*) \tag{7}$$

is defined for all Δy^*, zero or not, with $|(y^* + \Delta y^*) - y_0| < b^+$.

Lemma 1. The function $F(x, z, \Delta y^*)$ is jointly continuous in x, z, Δy^* in some region $\mathfrak{B}$ of $(x, z, \Delta y^*)$-space.

[1] We shall use the conventional notation $f_y(y, x) \equiv \frac{\partial}{\partial y} f(y, x)$ whenever convenient.

Proof. We first exhibit the region $\mathfrak{B}$ of the space of variables $\{x, z, \Delta y^*\}$ on which F is defined. By Theorem 5 of Chapter 4,

$$f(y + z\Delta y^*, x) = f(y+\Delta y, x) = f(\varphi(x, y^* + \Delta y^*, x^*), x)$$

and

$$f(y, x) = f(\varphi(x, y^*, x^*), x)$$

are defined for x such that $x_0 \leq x \leq c$ and $|y^* - y_0| \leq b^+$, $|x^* - x_0| \leq b^+$, $|y^* + \Delta y^* - y_0| \leq b^+$. Now let Δy^* be fixed. For each x such that $x_0 \leq x \leq c$ we can find a range of the z variable around $z = 0$ and an x neighborhood such that for x′ in this neighborhood, $|x - x'| < \varepsilon$ and $(\varphi(x) + z'\Delta y^*, x)$ is in $\mathfrak{A}$ for z′ in the range of z. By the Heine-Borel Theorem a finite number, N, of these neighborhoods cover the interval $x_0 \leq x \leq c$. For Δy^* fixed, the region $\mathfrak{B}$ can be constructed by taking for each value of x the minimum z range associated with the neighborhoods from this finite collection, N.

Hence there exists a region $\mathfrak{B}$ of three dimensional euclidean space on which the function $F(x, z, \Delta y^*)$ is defined. [Note that Equation (4) which we have used above holds only for $\Delta y^* \neq 0$. If $\Delta y^* = 0$, then F is given by Equation (6) and this function is given for $x_0 \leq x \leq c$ and for any z. Certainly, then $(\partial f/\partial y)\, z$ is defined for $\{x, z, 0\}$ in $\mathfrak{B}$.]

To prove the continuity of F we note that f(y, x) is a continuous function of x and hence jointly continuous in x, z and Δy^*. The product $z \cdot \Delta y^*$ is also jointly continuous in x, z, Δy^*. The quantity $y + z\Delta y^*$ is the sum of two continuous functions (y is a function of x, $z \cdot \Delta y^*$ a function of z and Δy^*) and hence is jointly continuous in x, z, Δy^*. The function f is jointly continuous in both arguments by hypothesis, and hence $f(y + z\Delta y^*, x)$ is jointly continuous in x, z, Δy^* since a continuous function of a continuous function is continuous. For Δy^* not zero, Equation (4) may be written

$$\frac{dz}{dx} = \frac{f(y + z\Delta y^*, x) - f(y, x)}{\Delta y^*} = F(x, z, \Delta y^*),$$

and hence for $\Delta y^* \neq 0$, F is jointly continuous in x, z, Δy^*.

Now let (x′, z′, 0) be some point in $\mathfrak{B}$. If $(x, y, \Delta y^*)$ is another point of $\mathfrak{B}$, then if $\Delta y^* \neq 0$, $z \neq 0$ and

$$F(x, z, \Delta y^*) - F(x', z', 0) = \frac{f(y + z\Delta y^*, x) - f(y,x)}{z\Delta y^*} z - f_y(y', x') \cdot z'$$
$$= f_y(y + \theta z\Delta y^*, x) \cdot z - f_y(y', x') \cdot z' \quad (8)$$

by the Law of the Mean for some θ between 0 and 1. One can also show that this formula is valid for $z = 0$ or $\Delta y^* = 0$, cf. Equation (6). Now consider

$$f_y(y + \alpha z, x)\, z$$

as a function of x, α and z. This is readily seen to be jointly continuous in x, z and α. Hence, given an $\varepsilon > 0$ there exists a $\delta > 0$ such that if

$$|\alpha| < \delta, \quad |x - x'| < \delta, \quad |z - z'| < \delta$$

then

$$|f_y(y + \alpha z, x)\, z - fy(y', x')z'| < \varepsilon.$$

Since $0 < \theta < 1$, this result and Equation (8) show that if $|\Delta y^*| < \delta$, $|x - x'| < \delta$, $|z - z'| < \delta$, then

$$|F(x, z, \Delta y^*) - F(x', z', 0)| < \varepsilon.$$

Thus F is continuous in x, z and Δy^* at $(x', z', 0)$.

We have therefore shown that $F(x, z, \Delta y^*)$ is jointly continuous in x, z and Δy^* in the region $\mathfrak{B}$, including points where $\Delta y^* = 0$. By our construction of $\mathfrak{B}$ it follows that:

Corollary. The set $\mathfrak{A}'$ of $\{\varphi(x) + z\Delta y^*, x\}$ for $\{x, z, \Delta y^*\}$ in $\mathfrak{B}$ is in $\mathfrak{A}$.

The set $\mathfrak{B}$ can be slightly restricted to say $\overline{\mathfrak{B}}$ such that the set $\overline{\mathfrak{A}}'$ corresponding to $\overline{\mathfrak{B}}$ is closed.

Lemma 2. $F(x, z, \Delta y^*)$ satisfies a Lipschitz condition on $\overline{\mathfrak{B}}$.

Proof. Here we must show that if $(x, z', \Delta y^*)$ and $(x, z'', \Delta y^*)$ are two points in $\overline{\mathfrak{B}}$, then there exists a constant L such that

$$|F(x, z', \Delta y^*) - F(x, z'', \Delta y^*)| \leq L\, |z' - z''|.$$

Case (i). $\Delta y^* \neq 0$. Then by Equation (4),

$$F(x, z', \Delta y^*) - F(x, z'', \Delta y^*)$$
$$= \frac{1}{\Delta y^*}\{[f(y + z'\Delta y^*, x) - f(y, x)] - [f(y + z''\Delta y^*, x) - f(y, x)]\}$$
$$= \frac{1}{\Delta y^*}[f(y + z'\Delta y^*, x) - f(y + z''\Delta y^*, x)].$$

Since $\frac{\partial}{\partial y}f(y, x)$ exists we may apply the Law of the Mean to the above difference, obtaining

$$\frac{1}{\Delta y^*}[f(y + z'\Delta y^*, x) - f(y + z''\Delta y^*, x)]$$
$$= \frac{1}{\Delta y^*} f_y(y^+, x)\,(z'\Delta y^* - z''\Delta y^*) = f_y(y^+, x)\,(z' - z'')$$

where (y^+, x) is a point intermediate between $(y + z'\Delta y^*, x)$ and $(y + z''\Delta y^*, x)$. We thus have

$$| F(x, z', \Delta y^*) - F(x, z'', \Delta y^*) | = | f_y(y^+, x) | \, | z' - z'' | .$$

Since $\partial f/\partial y$ is continuous on $\mathfrak{A}$ it is bounded on any closed bounded subset of $\mathfrak{A}$, in particular on $\overline{\mathfrak{A}}'$. Let L be this bound, $| \partial f/\partial y | \leq L$. Then

$$| F(x, z', \Delta y^*) - F(x, z'', \Delta y^*) | \leq L \, | z' - z'' |$$

and we have our Lipschitz condition on $\overline{\mathfrak{B}}$ for $\Delta y^* \neq 0$.

Case (ii). Suppose $\Delta y^* = 0$. Then

$$F(x, z', 0) - F(x, z'', 0) = f_y(y, x)\,(z' - z'')$$

and clearly F satisfies a Lipschitz condition on $\overline{\mathfrak{B}}$ in this case too.

2.3 Thus far we have shown that in the equation

$$\frac{dz}{dx} = F(x, z, \Delta y^*) \tag{7}$$

F is jointly continuous in the independent variable x, dependent variable z and parameter Δy^* on $\mathfrak{B}$ and satisfies a Lipschitz condition on $\overline{\mathfrak{B}}$. Furthermore, for $\Delta y^* \neq 0$, $z = \frac{\Delta y}{\Delta y^*}$.

We now wish to see what initial conditions z satisfies. To do this, consider Equations (1) and (2) with x replaced by x^*. Then

$$y(x^*) = \varphi(x^*, y^*, x^*) = y^*$$

by (ii) of the conclusion of Theorem 1. (Cf. Theorem 5 of Chapter 4.) Also

$$y(x^*) + \Delta y(x^*) = \varphi(x^*, y^* + \Delta y^*, x^*) = y^* + \Delta y^*.$$

Hence

$$\Delta y(x^*) = \Delta y^*$$

and

$$\frac{\Delta y(x^*)}{\Delta y^*} = 1 \quad \text{for} \quad \Delta y^* \neq 0.$$

We have thus shown that for $\Delta y^* \neq 0$, $z(x)$ is a solution of Equation (7) which passes through the point $(z, x) = (1, x^*)$. This also implies that for $x = x^*$, $\partial y/\partial y^* = 1$.

For each value of Δy^* (including $\Delta y^* = 0$) consider that solution of Equation (7) which passes through $(1, x^*)$. By the Lipschitz condition (of F on $\mathfrak{B}$) there is only one solution for each value of Δy^*, say

$$\psi(x) = \psi(x, 1, x^*, \Delta y^*).$$

Thus for $\Delta y^* \neq 0$ we have

$$z(x) = \psi(x, 1, x^*, \Delta y^*).$$

But from Theorem 8 of Chapter 4 with $n = 1$, $r = 1$, $\lambda_1 = \Delta y^*$, we conclude that $z(x)$ is continuous in Δy^* on $\mathfrak{B}$ (which includes points for which $\Delta y^* = 0$). Hence

$$\lim_{\Delta y^* \to 0} z(x) = \lim_{\Delta y^* \to 0} \psi(x, 1, x^*, \Delta y^*) = \psi(x, 1, x^*, 0)$$

(the last limit existing by the continuity of ψ). Since $z(x) = \Delta y/\Delta y^*$, $\lim\limits_{\Delta y^* \to 0} \dfrac{\Delta y}{\Delta y^*}$ exists and equals $\psi(x, 1, x^*, 0)$, that is

$$\frac{\partial y}{\partial y^*} = \psi(x, 1, x^*, 0)$$

exists.

This establishes the *existence* of $\partial y/\partial y^*$. Before proving the *continuity* of $\partial y/\partial y^*$ we note the following corollary:

Corollary. If $z = \partial y/\partial y^*$, then z satisfies the differential equation

$$\frac{dz}{dx} = f_y(\varphi(x, y^*, x^*), x) \cdot z$$

for $x_0 \leq x \leq c$, $|x^* - x_0| \leq b^+$, $|y^* - y_0| < b^+$ and at $x = x^*$, $z = 1$.

Proof. Since $z = \partial y/\partial y^* = \psi(x, 1, x^*, 0)$ satisfies

$$\frac{dz}{dx} = F(x, z, 0)$$

this means $\partial y/\partial y^*$ satisfies $dz/dx = (\partial f/\partial y)\, z$. The fact that $z = 1$ at $x = x^*$ is contained in the proof of the theorem.

2.4 To complete the proof of Theorem 1 we must establish the joint continuity of $z = \partial y/\partial y^*$ in x, y^* and x^*. By hypothesis $\partial f/\partial y$ exists and is jointly continuous in both variables. Also $\varphi(x, y^*, x^*)$ is jointly continuous in x, y^* and x^*. Hence, since a continuous function of a continuous function is continuous, $\partial f/\partial y$ is jointly continuous in x, y^*, x^*. Clearly, $(\partial f/\partial y)\, z$ is jointly continuous in x, y^*, x^* and z. We therefore have

$$\frac{dz}{dx} = G(z, x, y^*, x^*) = f_y(\varphi(x, y^*, x^*), x)\cdot z \tag{9}$$

where G is jointly continuous in z, x, y^* and x^*. Clearly G satisfies a Lipschitz condition for $L \geq |\, \partial f/\partial y \,|$. (In Equation (9) x^* and y^* are to be regarded as parameters, whereas in the original differential equation, they were initial condition parameters.) Since the hypothesis of Theorem 8 of Chapter 4 is fulfilled (we are considering the case $n = 1$, $r = 2$, $\lambda_1 = y^*$, $\lambda_2 = x^*$) we conclude that z is jointly continuous in x, y^* and x^*.

2.5 It is natural to ask why one does not use the Picard iterants to prove the above results, thus eliminating the lengthy limiting procedures. The answer is, that while it is possible to obtain a proof of the theorem in this manner, certain additional difficulties arise.

The Picard iterants $y^{(n)}(x)$ satisfy the equations

$$y^{(n)}(x) = y_0 + \int_{x_0}^{x} f(y^{(n-1)}(t), t)\; dt, \qquad n = 1, 2, \ldots .$$

If the first iterant (which would normally be y_0 itself) is differentiable relative to y_0, one can obtain

$$\frac{\partial y^{(n)}}{\partial y_0} = 1 + \int_{x_0}^{x} f_y(y^{(n-1)}(t), t)\, \frac{\partial y^{(n-1)}}{\partial y_0} dt$$

under the assumption that $f_y(y, x)$ exists and is continuous. Subtracting this equation for successive values of n will lead to

$$\frac{\partial}{\partial y_0}(y^{(n+1)} - y^{(n)}) = \int_{x_0}^{x} \left[f_y(y^{(n)}, t)\, \frac{\partial y^{(n)}}{\partial y_0} - f_y(y^{(n-1)}, t)\, \frac{\partial y^{(n-1)}}{\partial y_0} \right] dt$$

$$= \int_{x_0}^{x} \left\{ \left[f_y(y^{(n)}, t) \left(\frac{\partial y^{(n)}}{\partial y_0} - \frac{\partial y^{(n-1)}}{\partial y_0} \right) \right] + [f_y(y^{(n)}, t) - f_y(y^{(n-1)}, t)]\, \frac{\partial y^{(n-1)}}{\partial y_0} \right\} dt.$$

The first term in brackets in the integrand of the above equation is clearly amenable to the type of argument used in the Picard iteration process. To use the same type of argument on the second term requires that f_y itself satisfy a Lipschitz condition — which is a heavy assumption. Undoubtedly, somewhat more sophisticated arguments using the known convergence properties of the $y^{(n)}$ and the uniform continuity of f_y can be utilized to yield a result similar to the theorem given above; but such a discussion would have no real advantage over the above.

If on the other hand f_y does satisfy a Lipschitz condition, for example, if it were analytic, then this would be a convenient method of proof.

2.6 Notice that we have assumed the existence of a solution in the large, that is, the "running solution" of (iv) of the hypothesis of the above theorem defined on $[x_0, c]$. This will be done throughout the present chapter since it corresponds to the situation in which these theorems are normally employed. In any case, a solution in the small could be obtained by the use of previous existence theorems if (iv) were not hypothesized.

3. Differentiation with respect to y initial conditions for the case of n dependent variables

3.1 We shall now prove the analog of Theorem 1 for the case of n dependent variables. The theorem reads as follows.

Theorem 2. Hypothesis. (i) Let $f_i(y_1, \ldots, y_n, x)$, $i = 1, \ldots, n$ be n real valued functions of the $n + 1$ real variables $y_1, \ldots, y_n, x$ defined and continuous on a convex open region $\mathfrak{A}$ of $(n + 1)$-dimensional euclidean space.

(ii) Let $\partial f_i/\partial y_j$, $i, j = 1, \ldots, n$ exist and be jointly continuous in $y_1, \ldots, y_n, x$ on $\mathfrak{A}$.

(iii) Let $(y_{1,0}, \ldots, y_{n,0}, x_0)$ be a point of $\mathfrak{A}$.

(iv) There exists a solution $\{\varphi_1(x), \ldots, \varphi_n(x)\}$ to the system of differential equations

$$\frac{dy_i}{dx} = f_i(y_1, \ldots, y_n, x), \qquad i = 1, \ldots, n$$

defined on $x_0 \leq x \leq c$ with values $\{\varphi_1(x), \ldots, \varphi_n(x), x\}$ in $\mathfrak{A}$ and $y_{j,0} = \varphi_j(x_0)$.

Conclusion. There exists a constant $b^+ > 0$ and n functions

$$y_i = \varphi_i(x, y_1^*, \ldots, y_n^*, x^*)$$

defined on a region $\mathfrak{M}$, $x_0 \leq x \leq c$, $|y_i^* - y_{i,0}| \leq b^+$, $|x^* - x_0| \leq b^+$ such that:

(i) $\partial\varphi_i/\partial x = f_i(\varphi_1, \ldots, \varphi_n, x)$ for $x_0 \leq x \leq c$.

(ii) The $\varphi_i(x, y_1^*, \ldots, y_n^*, x^*)$ are jointly continuous in $x, y_1^*, \ldots, y_n^*, x^*$ on $\mathfrak{M}$.

(iii) $y_i^* = \varphi_i(x^*, y_1^*, \ldots, y_n^*, x^*)$.

(iv) The $\partial\varphi_i/\partial y_j^*$ exist and are continuous for $x_0 \leq x \leq c$, $|y_i^* - y_{i,0}| < b^+$, $|x^* - x_0| \leq b^+$.

We see that our hypothesis is identical with that of Theorem 5 of Chapter 4 except that (ii) is stronger. That is, the existence of the partials $\partial f_i/\partial y_j$ implies a Lipschitz condition. Hence all the conclusions of Theorem 5 of Chapter 4 are also true and are represented by (i), (ii) and (iii) of the present theorem. However, because of the stronger (ii) of the hypothesis we have the additional conclusion (iv), namely, the existence and continuity of the $\partial\varphi_i/\partial y_j^*$, i, j = 1, ..., n partials.

As in Theorem 1 of the present chapter, an open neighborhood of the y_i^* appears in (iv).

3.2 We shall suppose for simplicity in notation that j = 1, that is, we shall prove the existence and continuity of $\partial\varphi_i/\partial y_1^*$, i = 1, ..., n. Clearly, no loss of generality results, that is, our theorem will be true for *any* j, j = 1, ..., n.

Let $(y_1^*, \ldots, y_n^*, x^*)$ be a point such that $|y_i^* - y_{i,0}| < b^+$, $x_0 \leq x \leq c$. Consider an increment Δy_1^* taken so small that $y_1^* + \Delta y_1^*$ is in the b^+ neighborhood of $y_{1,0}$. We have then

$$y_i(x) = \varphi_i(x, y_1^*, \ldots, y_n^*, x^*) \quad \text{for} \quad x_0 \leq x \leq c \tag{10}$$

and by definition

$$y_i(x) + \Delta y_i(x) = \varphi_i(x, y_1^* + \Delta y_1^*, y_2^*, \ldots, y_n^*, x^*). \tag{11}$$

The arguments in the φ_i functions of Equations (10) and (11) differ only in the second argument, that is, $x, y_2^*, \ldots, y_n^*, x^*$ are the same in both equations. However, the y_i, i = 1, ..., n change even if we only change y_1^*, that is, the $\Delta y_i(x)$ for $i > 1$ are not necessarily zero.

Since both Equations (10) and (11) are solutions of the differential equations

$$\frac{dy_i}{dx} = f_i(y_1, \ldots, y_n, x), \qquad i = 1, \ldots, n$$

for $x_0 \leq x \leq c$, we have for this interval

$$\frac{dy_i}{dx} = f_i(y_1, \ldots, y_n, x)$$

and

$$\frac{d}{dx}(y_i + \Delta y_i) = \frac{dy_i}{dx} + \frac{d\Delta y_i}{dx} = f_i(y_1 + \Delta y_1, y_2 + \Delta y_2, \ldots, y_n + \Delta y_n, x),$$

and subtracting these equations for corresponding values of i,

$$\frac{d}{dx}\Delta y_i = f_i(y_1 + \Delta y_1, \ldots, y_n + \Delta y_n, x) - f_i(y_1, \ldots, y_n, x).$$

Adding and subtracting intermediate terms:

$$\begin{aligned}\frac{d}{dx}\Delta y_i = {} & f_i(y_1+\Delta y_1, \ldots, y_n+\Delta y_n, x) - f_i(y_1, y_2+\Delta y_2, \ldots, y_n+\Delta y_n, x) \\ & + f_i(y_1, y_2+\Delta y_2, \ldots, y_n+\Delta y_n, x) - f_i(y_1, y_2, y_3+\Delta y_3, \ldots, y_n+\Delta y_n, x) \\ & \cdot \quad \cdot \quad \cdot \quad \cdot \quad \cdot \quad \cdot \quad \cdot \quad \cdot \quad \cdot \quad \cdot \quad \cdot \\ & + f_i(y_1, y_2, \ldots, y_{n-1}, y_n + \Delta y_n, x) - f_i(y_1, \ldots, y_n, x).\end{aligned}$$

Dividing by Δy_1^*, we obtain

$$\frac{d}{dx}\left(\frac{\Delta y_i}{\Delta y_1^*}\right) = \sum_{k=1}^{n}\left[\frac{f_i(y_1, \ldots, y_{k-1}, y_k + \Delta y_k, \ldots, y_n + \Delta y_n, x)}{\Delta y_1^*} - \frac{f_i(y_1, \ldots, y_k, y_{k+1} + \Delta y_{k+1}, \ldots, y_n + \Delta y_n, x)}{\Delta y_1^*}\right] \qquad (12)$$

$$i = 1, \ldots, n.$$

If $\Delta y_k \neq 0$ multiply and divide the kth term on the right hand side of Equation (12) by Δy_k. If $\Delta y_k = 0$, the corresponding term is zero. Thus we obtain

$$\frac{d}{dx}\left(\frac{\Delta y_i}{\Delta y_1^*}\right) = \sum_{\substack{k=1 \\ \Delta y_k \neq 0}}^{n}\left[\frac{f_i(y_1, \ldots, y_{k-1}, y_k + \Delta y_k, \ldots, y_n + \Delta y_n, x)}{\Delta y_k} - \frac{f_i(y_1, \ldots, y_k, y_{k+1} + \Delta y_{k+1}, \ldots, y_n + \Delta y_n, x)}{\Delta y_k}\right]\frac{\Delta y_k}{\Delta y_1^*}. \qquad (13)$$

Now in each term of the brackets, replace Δy_k by $\frac{\Delta y_k}{\Delta y_1^*} \Delta y_1^*$, and in analogy with Theorem 1, let

$$z_k = \frac{\Delta y_k}{\Delta y_1^*}.$$

Thus Equation (13) becomes

$$\frac{dz_i}{dx} = \sum_{\substack{k=1 \\ \Delta y_k \neq 0}}^{n} \frac{f_i(y_1, \ldots, y_{k-1}, y_k + z_k \Delta y_1^*, \ldots, y_n + z_n \Delta y_1^*, x)}{z_k \Delta y_1^*} z_k$$
$$- \frac{f_i(y_1, \ldots, y_k, y_{k+1} + z_{k+1} \Delta y_1^*, \ldots, y_n + z_n \Delta y_1^*, x)}{z_k \Delta y_1^*} z_k \tag{14}$$

which we shall write more briefly as

$$\frac{dz_i}{dx} = F_i(x, z_1, \ldots, z_n, \Delta y_1^*) \quad \text{for} \quad \Delta y_1^* \neq 0. \tag{15}$$

In Equation (14) the y_i are *known* functions of x, that is, $y_i = \varphi_i(x, y_1^*, \ldots, y_n^*, x^*)$. Therefore, the F_i of Equation (15) are not to be regarded as functions of the y_i's. Equations (15) are to be regarded as differential equations on the dependent variables $z_1, \ldots, z_n$ with x as the independent variable and Δy_1^* as a parameter.

As earlier, we *define* $F_i(x, z_1, \ldots, z_n, \Delta y_1^*)$ for $\Delta y_1^* = 0$ as

$$F_i(x, z_1, \ldots, z_n, 0) = \sum_{k=1}^{n} \frac{\partial f_i}{\partial y_k} z_k. \tag{16}$$

Now $\partial f_i/\partial y_k$ exists and is jointly continuous in the y_i and x by (ii) of the hypothesis. Hence when y_i is replaced by a continuous function of x, $y_i = \varphi_i(x, y_1^*, \ldots, y_n^*, x^*)$, $\partial f_i/\partial y_k$ becomes a continuous function of x. Clearly $\Sigma (\partial f_i/\partial y_k)z_k$ is jointly continuous in x and the z_k. Hence the differential equation on z_i,

$$\frac{dz_i}{dx} = F_i(x, z_1, \ldots, z_n, \Delta y_1^*)$$

is defined for all Δy_1^*, zero or not, for which $|(y_1^* + \Delta y_1^*) - y_{1,0}| < b^+$.

Lemma 3. The functions $F_i(x, z_1, \ldots, z_n, \Delta y_1^*)$ are jointly continuous in all $n + 2$ variables in some convex open region $\mathfrak{B}$ of $(x, z_1, \ldots, z_n, \Delta y_1^*)$ — space including points where $\Delta y_1^* = 0$.

The proof of this lemma is, except for complications of notation, the same as the proof of Lemma 1 and will be omitted.

3.3 We now wish to show that the $F_i(x, z_1, \ldots, z_n, \Delta y_1^*)$ satisfy a Lipschitz condition on $\overline{\mathfrak{B}}$, a closed convex subset of $\mathfrak{B}$. We must show that if $(x, z_1', \ldots, z_n, \overline{\Delta y_1^*})$ and $x, z_1'', \ldots, z_n'', \Delta y_1^*)$ are two points in $\overline{\mathfrak{B}}$, then there exists a constant L such that

$$| F_i(x, z_1', \ldots, z_n', \Delta y_1^*) - F_i(x, z_1'', \ldots, z_n'', \Delta y_1^*) | \leq L \sum_{k=1}^{n} | z_k' - z_k'' | .$$

Case (i). $\Delta y_1^* \neq 0$.

By definition of the F_i

$$\begin{aligned} F_i(x, z_1', \ldots, z_n', \Delta y_1^*) &- F_i(x, z_1'', \ldots, z_n'', \Delta y_1^*) \\ &= \frac{1}{\Delta y_1}\{[f_i(y_1 + z_1' \Delta y_1^*, \ldots, y_n + z_n' \Delta y_1^*, x) - f_i(y_1, \ldots, y_n, x)] \\ &\quad - [f_i(y_1 + z_1'' \Delta y_1^*, \ldots, y_n + z_n'' \Delta y_1^*, x) - f_i(y_1, \ldots, y_n, x)]\} \\ &= \frac{1}{\Delta y_1^*} [f_i(y_1 + z_1' \Delta y_1^*, \ldots, y_n + z_n' \Delta y_1^*, x) \\ &\qquad - f_i(y_1 + z_1'' \Delta y_1^*, \ldots, y_n + z_n'' \Delta y_1^*, x)]. \end{aligned}$$

Since the $\partial f_i/\partial y_j$ exist by hypothesis we may apply the Law of the Mean in n variables to the above difference obtaining

$$\begin{aligned} F_i(x, z_1', \ldots, z_n', \Delta y_1^*) &- F_i(x, z_1'' . \ldots, z_n'', \Delta y_1^*) \\ &= \frac{1}{\Delta y_1^*}\left[\sum_{k=1}^{n} \left(\frac{\partial f_i}{\partial y_k}\right)(z_k' \Delta y_1^* - z_k'' \Delta y_1^*)\right] \end{aligned} \tag{17}$$

where $\partial f_i/\partial y_k$ is evaluated at the point $(x, z_1^+, \ldots, z_n^+, \Delta y_1^*)$, z_k^+ being a point intermediate between z_k' and z_k''. The $\partial f_1/\partial y_k$ are continuous on the closed set $\overline{\mathfrak{B}}$ and hence are bounded. Therefore, as in Lemma 2, Equation (17) clearly implies a Lipschitz condition.

Case (ii). $\Delta y_1^* = 0$.

Again, by definition [Equation (16)],

$$F_i(x, z_1, \ldots, z_n, 0) = \sum_{k=1}^{n} \frac{\partial f_i}{\partial y_k} z_k$$

and

$$F_i(x, z_1', \ldots, z_n', 0) - F_i(x, z_1'', \ldots, z_n'', 0) = \sum_{k=1}^{n} \left(\frac{\partial f_i}{\partial y_k}\right)(z_k' - z_k'')$$

which, as above, clearly implies a Lipschitz condition.

We have therefore proved:

Lemma 4. $F_i(x, z_1, \ldots, z_n, \Delta y_1^*)$ satisfies a Lipschitz condition on $\overline{\mathfrak{B}}$. If L is such that $|\partial f_i/\partial y_k| \leq L$ on $\overline{\mathfrak{B}}$, then L may be used as the constant of the Lipschitz condition.

3.4 Thus we have the following results for the equations

$$\frac{dz_i}{dx} = F_i(x, z_1, \ldots, z_n, \Delta y_1^*), \qquad i = 1, \ldots, n. \tag{18}$$

The F_i are jointly continuous in the independent variable x, dependent variables z_i and parameter Δy_1^* on $\mathfrak{B}$, and satisfy a Lipschitz condition on $\overline{\mathfrak{B}}$. Furthermore, for $\Delta y_1^* \neq 0$, $z_i = \Delta y_i/\Delta y_1^*$ satisfies this equation.

We now wish to see what initial conditions the z_i satisfy. To do this, consider Equations (10) and (11) with x replaced by x^* and $i = 1$. Then

$$y_1(x^*) = \varphi_1(x^*, y_1^*, \ldots, y_n^*, x^*) \equiv y_1^*$$

$$y_1(x^*) + \Delta y_1(x^*) = \varphi_1(x^*, y_1^* + \Delta y_1^*, y_2^*, \ldots, y_n^*, x^*) \equiv y_1^* + \Delta y_1^*.$$

Therefore

$$\Delta y_1(x^*) = \Delta y_1^*$$

and

$$\frac{\Delta y_1(x^*)}{\Delta y_1^*} = 1.$$

Similarly, for $i > 1$,

$$y_i(x^*) \equiv y_i^*$$

and

$$y_i(x^*) + \Delta y_i(x^*) = y_i^* + \Delta y_i^* = y_i^*$$

(since $\Delta y_i^* = 0$, $i = 2, \ldots, n$). Hence

$$\frac{\Delta y_i(x^*)}{\Delta y_1^*} = 0, \qquad i = 2, \ldots, n.$$

We have thus shown that for $\Delta y_1^* \neq 0$, $z_i(x)$, $i = 1, \ldots, n$ is a solution of Equation (18) which passes through the point

$$(z_1, \ldots, z_n, x) = (1, 0, \ldots, 0, x^*).$$

This also yields

$$\frac{\partial y_i^*}{\partial y_1^*} = \delta_{i1}, \qquad i = 1, \ldots, n.$$

For each value of Δy_1^* (including $\Delta y_1^* = 0$) consider that solution of Equation (18) which passes through $(1, 0, \ldots, 0, x^*)$. By the Lipschitz condition of F_i on $\overline{\mathfrak{B}}$ there is only one solution for each value of Δy_1^*, say

$$\psi_i(x) = \psi_i(x, \Delta y_1^*).$$

Thus for $\Delta y_1^* \neq 0$ we have

$$z_i(x) = \psi_i(x, \Delta y_1^*).$$

But from Theorem 8 of Chapter 4 with $r = 1$, $\lambda_1 = \Delta y_1^*$ we conclude that $z_i(x)$ is continuous in Δy_1^* on $\overline{\mathfrak{B}}$ (which includes points for which $\Delta y_1^* = 0$). Hence

$$\lim_{\Delta y_1^* \to 0} z_i(x) = \lim_{\Delta y_1 \to 0} \psi_i(x, \Delta y_1^*) = \psi_i(x, 0)$$

(the last limit existing by the continuity of the ψ_i). Since $z_i(x) = \Delta y_i/\Delta y_1^*$,

$$\lim_{\Delta y_1^* \to 0} \frac{\Delta y_i}{\Delta y_1^*}$$

exists and equals $\psi_i(x, 0)$. That is,

$$\partial y_i/\partial y_1^* = \psi_i(x, 0)$$

exists and at $x = x^*$, $\partial y_i/\partial y_1^* = \delta_{i1}$.

This establishes the *existence* of $\partial y_i/\partial y_1^*$. To complete the proof of Theorem 2 we must establish the joint continuity of $z_i = \partial y_i/\partial y_1^*$ in $x, y_1^*, \ldots, y_n^*, x^*$. The proof of this fact follows that given in Section 2.4 for the case $n = 1$.

3.5 If in the above proof we had considered Δy_k^* instead of Δy_1^* (with $\Delta y_j^* = 0$, $j \neq k$) then it would follow:

Corollary. The

$$z_{ik} = \frac{\partial \varphi_i}{\partial y_k}, \qquad i = 1, \ldots, n \tag{20}$$

exist and are continuous and satisfy

$$\frac{d}{dx} z_{ik} = \sum_{j=1}^{n} \frac{\partial f_i}{\partial y_j} z_{jk}, \qquad i, k = 1, \ldots, n \tag{21}$$

with the initial condition

$$z_{ik}(x^*) = \delta_{ik}. \tag{22}$$

Note that Equation (21) is a system of *linear* differential equations on the unknown functions $z_{ik} = \partial y_i/\partial y_k^*$. (Linear equations will be treated in detail in Chapter 6.)

4. Differentiation with respect to parameters

4.1 If the initial differential equation contains parameters $\lambda_1, \ldots, \lambda_r$, that is,

$$\frac{dy_i}{dx} = f_i(y_1, \ldots, y_n, x, \lambda_1, \ldots, \lambda_r),$$

then under suitable conditions we can show that the solutions

$$y_i = \varphi_i (x, \lambda_1, \ldots, \lambda_r)$$

have continuous derivatives with respect to the parameters λ_k. The pattern of proof closely follows that for y initial conditions. Precisely formulated, the theorem reads as follows.

Theorem 3. Hypothesis. (i) Let $f_i(y_1, \ldots, y_n, x, \lambda_1, \ldots, \lambda_r)$ be n real valued functions of the $n + 1 + r$ real variables $y_1, \ldots, y_n, x, \lambda_1, \ldots, \lambda_r$ defined and continuous on a convex open region $\mathfrak{A}$ of $(n + 1 + r)$-dimensional euclidean space.

(ii) Let $\partial f_i/\partial y_j$ and $\partial f_i/\partial \lambda_k$, $i, j = 1, \ldots, n$, $k = 1, \ldots, r$ exist and be jointly continuous in $y_1, \ldots, y_n, x, \lambda_1, \ldots, \lambda_r$ on $\mathfrak{A}$.

(iii) Let $(y_{1,0}, \ldots, y_{n,0}, x_0, \lambda_{1,0}, \ldots, \lambda_{r,0})$ be a point in $\mathfrak{A}$.

(iv) There exists a solution $\{\varphi_1(x), \ldots, \varphi_n(x)\}$ to the system of differential equations

$$\frac{dy_i}{dx} = f_i(y_1, \ldots, y_n, x, \lambda_{1,0}, \ldots, \lambda_{r,0}), \qquad i = 1, \ldots, n$$

for $x_0 \leq x \leq c$ with values $\{\varphi_1(x), \ldots, \varphi_n(x), x, \lambda_{1,0}, \ldots, \lambda_{r,0}\}$ in $\mathfrak{A}$ and $y_{j,0} = \varphi_j(x_0)$.

Conclusion. There exists a constant $a > 0$ and n functions

$$y_i = \varphi_i(x, \lambda_1, \ldots, \lambda_r)$$

defined on a region $\mathfrak{M}$, $x_0 \leq x \leq c$, $| \lambda_i - \lambda_{i,0} | \leq a$, such that

(i) $\partial\varphi_i/\partial x = f_i(\varphi_1, \ldots, \varphi_n, x)$ for $x_0 \leq x \leq c$, $\qquad i = 1, \ldots, n$.

(ii) The φ_i are jointly continuous in the $r + 1$ variables $x, \lambda_1, \ldots, \lambda_r$ on $\mathfrak{M}$, $\quad i = 1, \ldots, n$.

(iii) $y_{i,0} = \varphi_i(x_0, \lambda_1, \ldots, \lambda_r)$.

(iv) $\partial\varphi_j/\partial\lambda_k$ exists and is continuous for $x_0 \leq x \leq c$, $|\lambda_{k,0} - \lambda_k| < a$.

4.2 Since the hypotheses of this theorem are stronger than those of Theorem 8 of Chapter 4, the conclusions of this theorem follow. However, the existence and continuity of the partials with respect to the λ's enable us to deduce the additional conclusions stated in (iv).

To prove Theorem 3, we shall, for simplicity, assume k = 1, that is, prove the existence and continuity of $\partial\varphi_j/\partial\lambda_1$. Clearly, no loss of generality will result, and our theorem will also be true for λ_k, $k > 1$. Consider a λ point, say λ^*, such that $|\lambda_{k,0} - \lambda_k^*| < a$. Let $\Delta\lambda_1^*$ be an increment on λ_1^* such that $|\lambda_{k,0} - (\lambda_k^* + \Delta\lambda_k^*)| \leqq a$. We have, then

$$y_i(x) = \varphi_i(x, \lambda_1^*, \ldots, \lambda_r^*) \quad \text{for} \quad x_0 \leqq x \leqq c$$

and by definition

$$y_i(x) + \Delta y_i(x) = \varphi_i(x, \lambda_1^* + \Delta\lambda_1^*, \lambda_2^*, \ldots, \lambda_r^*) \quad \text{for} \quad x_0 \leqq x \leqq c.$$

Both these functions are solutions of the differential equation

$$\frac{dy_i}{dx} = f_i(y_1, \ldots, y_n, x, \lambda_1, \ldots, \lambda_r)$$

for the x interval $x_0 \leqq x \leqq c$. Hence

$$\frac{dy_i}{dx} = f_i(y_1, \ldots, y_n, x, \lambda_1^*, \ldots, \lambda_r^*)$$

$$\frac{dy_i}{dx} + \frac{d}{dx}\Delta y_i = f_i(y_1 + \Delta y_1, \ldots, y_n + \Delta y_n, x, \lambda_1^* + \Delta\lambda_1^*, \lambda_2^*, \ldots, \lambda_r^*).$$

Subtracting

$$\frac{d}{dx}\Delta y_i = f_i(y_1 + \Delta y_1, \ldots, y_n + \Delta y_n, x, \lambda_1^* + \Delta\lambda_1^*, \lambda_2^*, \ldots, \lambda_r^*) - f_i(y_1, \ldots, y_n, x, \lambda_1^*, \ldots, \lambda_r^*).$$

We shall assume that $\Delta y_j \neq 0$ for $\Delta\lambda_1^* \neq 0$. If this assumption is not made, certain expressions in the following will be zero; but the proof is quite similar with a more complicated notation.

Performing the same manipulations as in Theorem 2 using $\Delta\lambda_1^*$ in place of Δy_1^* we arrive at

$$\frac{dz_i}{dx} = \sum_{k=1}^{n} \frac{f_i(y_1, \ldots, y_{k-1}, y_k + z_k\Delta\lambda_1^*, \ldots, y_n + z_n\Delta\lambda_1^*, x, \lambda_1^* + \Delta\lambda_1^*, \ldots, \lambda_r^*)}{z_k\,\Delta\lambda_1^*} z_k$$
$$- \frac{f_i(y_1, \ldots, y_k, y_{k+1} + z_{k+1}\Delta\lambda_1^*, \ldots, y_n + z_n\Delta\lambda_1^*, x, \lambda_1^* + \Delta\lambda_1^*, \ldots, \lambda_r^*)}{z_k\,\Delta\lambda_1^*} z_k$$
$$+ \frac{f_i(y_1, \ldots, y_n, x, \lambda_1^* + \Delta\lambda_1^*, \lambda_2^*, \ldots, \lambda_r^*) - f_i(y_1, \ldots, y_n, x, \lambda_1^*, \ldots, \lambda_r^*)}{\Delta\lambda_1^*}$$

where

$$z_k = \frac{\Delta y_k}{\Delta\lambda_1^*}.$$

We shall write this as

$$\frac{dz_i}{dx} = G_i(x, z_1, \ldots, z_n, \Delta\lambda_1^*), \qquad \Delta\lambda_1^* \neq 0, \qquad i = 1, \ldots, n$$

and define G_i for $\Delta\lambda_1^* = 0$ as

$$G_i(x, z_1, \ldots, z_n, 0) = \sum_{k=1}^{n} \frac{\partial f_i}{\partial y_k} z_k + \frac{\partial f_i}{\partial \lambda_1^*}.$$

We can show as before that the G_i functions are jointly continuous in all $n + 2$ variables in some region $\mathfrak{B}$ of $(x, z_1, \ldots, z_n, \Delta\lambda_1^*)$-space (including points where $\Delta\lambda_1^* = 0$) and satisfy a Lipschitz condition on a closed $\overline{\mathfrak{B}} < \mathfrak{B}$.

4.3 The initial conditions satisfied by the z_i differential equations are

$$z_i(x_0) = 0, \qquad i = 1, \ldots, n.$$

The proof of the existence and continuity of the $\partial y_i/\partial\lambda_1^*$ follows as in the last two theorems. The corresponding corollary is:

Corollary. If $z_i = \partial y_i/\partial\lambda_1^*$, the z_i satisfy the differential equations

$$\frac{dz_i}{dx} = \sum_{k=1}^{n} \frac{\partial f_i}{\partial y_k} z_k + \frac{\partial f_i}{\partial \lambda_1^*}$$

for $x_0 \leq x \leq c$, $|\lambda_{i,0} - \lambda_i^*| < a$ and at $x = x_0$, $z_k(x_0) = 0$.

4.4 The above theorem can be extended to include y initial conditions as well. (Cf. Theorem 8 of Chapter 4.) We state (but shall not prove) this theorem.

Theorem 4. Hypothesis. (i) Let $f_i(y_1, \ldots, y_n, x, \lambda_1, \ldots, \lambda_r)$ be n real valued functions of the $n + 1 + r$ real variables $y_1, \ldots, y_n, x, \lambda_1, \ldots, \lambda_r$ defined and continuous on a convex open region $\mathfrak{A}$ of $(n + 1 + r)$-dimensional euclidean space.

(ii) Let $\partial f_i/\partial y_j$ and $\partial f_i/\partial \lambda_k$, i, j = 1, . . ., n, k = 1, . . ., r exist and be jointly continuous in $y_1, \ldots, y_n, x, \lambda_1, \ldots, \lambda_r$ on $\mathfrak{A}$.

(iii) Let $(y_{1,0}, \ldots, y_{n,0}, x_0, \lambda_{1,0}, \ldots, \lambda_{r,0})$ be a point of $\mathfrak{A}$.

(iv) There exists a solution $\varphi_j(x)$, j = 1, . . ., n of the differential equations

$$\frac{dy_i}{dx} = f_i(y_1, \ldots, y_n, x, \lambda_{1,0}, \ldots, \lambda_{r,0})$$

i = 1, . . ., n defined on $x_0 \leq x \leq c$, with values $\{\varphi_1(x), \ldots, \varphi_n(x), x, \lambda_{1,0}, \ldots, \lambda_{r,0}\}$ in $\mathfrak{A}$ and $y_{j,0} = \varphi_j(x_0)$.

Conclusion. There exist constants $b^+ > 0$ and $a > 0$ and n functions

$$y_i = \varphi_i(x, y_1^*, \ldots, y_n^*, x^*, \lambda_1, \ldots, \lambda_r)$$

defined on a region $\mathfrak{M}$, $x_0 \leq x \leq c$, $|y_i^* - y_{i,0}| \leq b^+$, $|x^* - x_0| \leq b^+$, $|\lambda_i - \lambda_{i,0}| \leq a$ such that

(i) $\partial\varphi_i/\partial x = f_i(\varphi_1, \ldots, \varphi_n, x, \lambda_1, \ldots, \lambda_r)$ for $x_0 \leq x \leq c$.

(ii) The $\varphi_i(x, y_1^*, \ldots, y_n^*, x^*, \lambda_1, \ldots, \lambda_r)$ are jointly continuous in all n + r + 2 variables on $\mathfrak{M}$.

(iii) $y_i^* = \varphi_i(x^*, y_1^*, \ldots, y_n^*, x^*, \lambda_1, \ldots, \lambda_r)$.

(iv) $\partial\varphi_j/\partial y_i^*$ and $\partial\varphi_j/\partial\lambda_k^*$ exist and are continuous for $x_0 \leq x \leq c$, $|y_i^* - y_{i,0}| < b^+$, $|x^* - x_0| \leq b^+$, $|\lambda_k - \lambda_{k,0}| < a$.

5. Differentiation with respect to x initial condition

5.1 Having now considered y initial conditions and parameters we turn to the x initial condition.

Theorem 5. Hypothesis. (i) Let $f_i(y_1, \ldots, y_n, x)$, i = 1, . . ., n be n real valued functions of the n + 1 real variables $y_1, \ldots, y_n, x$ defined and continuous on a convex open region $\mathfrak{A}$ of (n + 1)-dimensional euclidean space.

(ii) Let $\partial f_i/\partial y_j$, i, j = 1, . . ., n exist and be continuous in $y_1, \ldots, y_n, x$ on $\mathfrak{A}$.

(iii) Let $(y_{1,0}, \ldots, y_{n,0}, x_0)$ be a point of $\mathfrak{A}$.

(iv) There exists a solution $y_j = \varphi_j(x)$, j = 1, . . ., n of the differential equations

$$\frac{dy_i}{dx} = f_i(y_1, \ldots, y_n, x), \qquad i = 1, \ldots, n$$

defined on $x_0 \leq x \leq c$ with values $\{\varphi_1(x), \ldots, \varphi_n(x), x\}$ in $\mathfrak{A}$ and $y_{j,0} = \varphi_j(x_0)$.

Conclusion. There exists a constant $b^+ > 0$ and n functions

$$y_i = \varphi_i(x, y_1^*, \ldots, y_n^*, x^*)$$

defined on a region $\mathfrak{M}$, $x_0 \leq x \leq c$, $|y_i^* - y_{i,0}| \leq b^+$, $|x^* - x_0| \leq b^+$ such that

(i) $\partial\varphi_i/\partial x = f_i(\varphi_1, \ldots, \varphi_n, x)$ for $x_0 \leq x \leq c$.

(ii) The $\varphi_i(x, y_1^*, \ldots, y_n^*, x^*)$ are jointly continuous in $x, y_1^*, \ldots, y_n^*, x^*$ on $\mathfrak{M}$.

(iii) $y_i^* = \varphi_i(x^*, y_1^*, \ldots, y_n^*, x^*)$.

(iv) $\partial\varphi_i/\partial x^*$ exists and is continuous for $x_0 \leq x \leq c$, $|y_i^* - y_{i,0}| < b^+$, $|x^* - x_0| < b^+$.

Comparing with Theorem 5 of Chapter 4, we see that (i) — (iii) of the conclusions follow immediately. We need therefore but to prove (iv).

5.2 Let $(y_1^*, \ldots, y_n^*, x^*)$ be an interior point of the b^+ interval, that is $|y_i^* - y_{i,0}| < b^+$, $|x^* - x_0| < b^+$. Consider an increment Δx^* on x^*. This will result in an increment Δy_i on the y_i. Let Δx^* be so small that $|x^* + \Delta x^* - x_0| \leq b^+$.

We have then

$$y_i(x) = \varphi_i(x, y_1^*, \ldots, y_n^*, x^*) \quad \text{for} \quad x_0 \leq x \leq c \tag{23}$$

and by definition

$$y_i(x) + \Delta y_i(x) = \varphi_i(x, y_1^*, y_2^*, \ldots, y_n^*, x^* + \Delta x^*). \tag{24}$$

Since both Equations (23) and (24) are solutions of the differential equation

$$\frac{dy_i}{dx} = f_i(y_1, \ldots, y_n, x), \qquad i = 1, \ldots, n$$

for $x_0 \leq x \leq c$, we have, for this interval

$$\frac{dy_i}{dx} = f_i(y_1, \ldots, y_n, x)$$

and

$$\frac{d}{dx}(y_i + \Delta y_i) = \frac{dy_i}{dx} + \frac{d\Delta y_i}{dx} = f_i(y_1 + \Delta y_1, y_2 + \Delta y_2, \ldots, y_n + \Delta y_n, x).$$

Subtract these two equations for corresponding values of i,

$$\frac{d}{dx}\Delta y_i = f_i(y_1 + \Delta y_1, \ldots, y_n + \Delta y_n, x) - f_i(y_1, \ldots, y_n, x).$$

We shall make the assumption that $\Delta y_i \neq 0$ if $\Delta x^* \neq 0$. Again this condition simplifies the notation since we need not explicitly describe certain zero terms. Now adding and subtracting intermediate terms from the above equation and dividing by Δx^*, we obtain

$$\frac{d}{dx}\left(\frac{\Delta y_i}{\Delta x^*}\right) = \sum_{k=1}^{n}\left[\frac{f_i(y_1, \ldots, y_{k-1}, y_k + \Delta y_k, \ldots, y_n + \Delta y_n, x)}{\Delta x^*} - \frac{f_i(y_1, \ldots, y_k, y_{k+1} + \Delta y_{k+1}, \ldots, y_n + \Delta y_n, x)}{\Delta x^*}\right]. \tag{25}$$

On the right hand side of Equation (25) multiply and divide the kth term by Δy_k, and in each term of the brackets, replace Δy_k by $(\Delta y_k/\Delta x^*)\Delta x^*$. Then if we let

$$u_k = \frac{\Delta y_k}{\Delta x^*},$$

Equation (25) becomes

$$\frac{du_i}{dx} = \sum_{k=1}^{n}\left[\frac{f_i(y_1, \ldots, y_{k-1}, y_k + u_k\,\Delta x^*, \ldots, y_n + u_n\,\Delta x^*, x)}{u_k\,\Delta x^*} - \frac{f_i(y_1, \ldots, y_k, y_{k+1} + u_{k+1}\,\Delta x^*, \ldots, y_n + u_n\,\Delta x^*, x)}{u_k\,\Delta x^*}\right] u_k \tag{26}$$

which we shall write more briefly as

$$\frac{du_i}{dx} = F_i(x, u_1, \ldots, u_n, \Delta x^*) \quad \text{for} \quad \Delta x^* \neq 0. \tag{27}$$

In the above equations we have assumed $\Delta x^* \neq 0$. If $\Delta x^* = 0$, we define F_i, as before, as:

$$F_i(x, u_1, \ldots, u_n, 0) = \sum_{k=1}^{n} \frac{\partial f_i}{\partial y_k} u_k. \tag{28}$$

Clearly $\Sigma\,(\partial f_i/\partial y_k)\,u_k$ is jointly continuous in x and the u_k and hence the differential equation on u_i,

$$\frac{du_i}{dx} = F_i(x, u_1, \ldots, u_n, \Delta x^*)$$

is defined for all Δx^*, zero or not.

We can show as in Theorem 2 that F_i is jointly continuous in some region $\mathfrak{B}$ of $(x, u_1, \ldots, u_n, \Delta x^*)$-space, $x_0 \leq x \leq c$ and satisfies a Lipschitz condition on $\bar{\mathfrak{B}}$, any closed subset of $\mathfrak{B}$.

5.3 Since the u_i above satisfy the same equation as $z_i = \dfrac{\partial \varphi_i}{\partial y_i^*}$ of Theorem 2, we can conclude that u_i and z_i differ only by initial conditions. In particular, we have that the $\partial\varphi_i/\partial x^*$ are continuous and by the corollary to Theorem 2

$$\frac{du_i}{dx} = \sum_{k=1}^{n} \frac{\partial f_i}{\partial y_k} u_k$$

where

$$u_i = \frac{\partial y_i}{\partial x^*}.$$

5.4 It is not necessary for the proof of the theorem, but it is interesting from both a practical and theoretical point of view to determine what initial conditions are satisfied by the u_i at $x = x^*$.

Now

$$y_i = \varphi_i(x, y_1^*, \ldots, y_n^*, x^*)$$

and by Theorems 2 and 5, y_i has a continuous derivative with respect to the y_i^*, x^* and x with

$$\frac{\partial y_i}{\partial x} = f_i(y_1, \ldots, y_n, x). \tag{29}$$

The functions φ_i are such that we have identically

$$y_i^* = \varphi_i(x^*, y_1^*, \ldots, y_n^*, x^*).$$

Since the $y_1^*, \ldots, y_n^*, x^*$ are independent variables

$$0 = \frac{\partial}{\partial x} \varphi_i(x^*, y_1^*, \ldots, y_n^*, x^*) + \frac{\partial}{\partial x^*} \varphi_i(x^*, y_1^*, \ldots, y_n^*, x^*).$$

From Equation (29) we can now infer that the initial conditions for the u_i are given by

$$u_i(x^*, y_1^*, \ldots, y_n^*, x^*) = \frac{\partial \varphi_i}{\partial x^*} = -\frac{\partial \varphi_i}{\partial x} = -f_i(y_1^*, \ldots, y_n^*, x^*).$$

6. Higher derivatives with respect to y initial conditions

6.1 In Theorem 2 we found that conditions sufficient to insure the existence and continuity of the derivatives of solutions with respect to y initial conditions were the existence and continuity of the partials of the original functions with respect to the dependent variables. We now wish to see what additional conditions are necessary for the existence and continuity of the *second* partials. That is, if

$$\frac{dy_i}{dx} = f_i(y_1, \ldots, y_n, x), \qquad i = 1, \ldots, n \tag{30}$$

with solution

$$y_i = \varphi_i(x, y_1^*, \ldots, y_n^*, x^*)$$

we wish to determine what restrictions in addition to those already given in the hypothesis of Theorem 2 we must place on the f_i functions in order that the $\partial^2 y_i/\partial y_j^* \partial y_k^*$ exist and be continuous.

From the corollary to Theorem 2 and the theorem itself we have

$$\frac{d}{dx} z_{ik} = \sum_{j=1}^{n} \frac{\partial f_i}{\partial y_j}(y_1, \ldots, y_n, x)\; z_{jk}(x) \tag{31}$$

where

$$z_{jk}(x) = \frac{\partial y_j}{\partial y_k^*} \tag{32}$$

and

$$z_{jk}(x^*) = \delta_{jk}.$$

In Equations (31), the dependent variables are the z_{ij} and the $y_1, \ldots, y_n$ are to be regarded as differentiable functions of x and $y_1^*, \ldots, y_n^*$. If in addition the $\partial f_i/\partial y_j$ are differentiable as functions of $y_1, \ldots, y_n$, then the system of Equation (31) is such that Theorem 3 above is applicable with the parameters λ_k replaced by the variables $y_1^*, \ldots, y_n^*$. This would then yield the existence and continuity properties of

$$w_{ijk} = \frac{\partial z_{ij}}{\partial y_k^*} = \frac{\partial^2 y_i}{\partial y_k^* \partial y_j^*}$$

and the fact that they satisfy:

$$\frac{d}{dx} w_{ijk} = \sum_{h=1}^{n} \sum_{\lambda=1}^{n} \frac{\partial^2 f_i}{\partial y_\lambda \partial y_h} z_{hj}\, z_{\lambda k} + \sum_{h=1}^{n} \frac{\partial f_i}{\partial y_h} w_{hjk}. \tag{33}$$

6.2 Hence the existence and continuity of the second partials with respect to y initial conditions require the existence and continuity of the second partials of the f_i functions. Generalizing, we may state:

Theorem 6. Hypothesis. (i) The entire hypothesis of Theorem 2.

(ii) Let $\dfrac{\partial^s f_i}{\partial y_1^{s_1} \partial y_2^{s_2} \dots \partial y_n^{s_n}}$ exist and be jointly continuous in $y_1, \dots,$ y_n, x for all $s_i \geq 0$ with $\sum_{i=1}^{n} s_i = s$.

Conclusion. (i) The entire conclusions of Theorem 2.

(ii) The $\dfrac{\partial^s \varphi_j}{\partial y_1^{*s_1} \partial y_2^{*s_2} \dots \partial y_n^{*s_n}}$ exist and are continuous for $x_0 \leq x \leq c$, $|y_i^* - y_{i,0}| < b^+$, $|x^* - x_0| \leq b^+$.

6.3 We see from Equation (31) that the $z_{ik} = \dfrac{\partial y_i}{\partial y_k^*}$ satisfy a homogeneous linear differential equation. Also, from Equation (33) we conclude that the $\partial^2 y_i / \partial y^* \partial y_k^*$ satisfy a non-homogeneous linear differential equation with the same homogeneous part as the z_{ik}. This is easily seen to be the general case and we have:

Corollary. All the partials $\dfrac{\partial^s \varphi_j}{\partial y_1^{*s_1} \dots \partial y_n^{*s_n}}$ satisfy a linear differential equation with the same homogeneous part.

We also note that the initial conditions for all higher partials are zero, for example,

$$\frac{\partial^2 y_i}{\partial y_j^* \partial y_k^*}$$

evaluated at $x = x^*$ equals zero since $\partial y_i / \partial y_k^* = \delta_{ik}$. So for the first partials z_{ik} we have a homogeneous linear differential equation with non-homogeneous initial conditions, while for the higher derivatives we have a non-homogeneous linear differential equation with homogeneous initial conditions.

7. Higher derivatives with respect to parameters

7.1 The same line of reasoning applies to differential equations with parameters.

Theorem 7. Hypothesis. (i) The entire hypothesis of Theorem 3.

(ii) Let $\dfrac{\partial^s f_i}{\partial y_1^{s_1} \ldots \partial y_n^{s_n} \partial \lambda_1^{s_{n+1}} \ldots \partial \lambda_r^{s_{n+r}}}$ exist and be jointly continuous in $y_1, \ldots, y_n, x, \lambda_1, \ldots, \lambda_r$ for all $s_i \geq 0$ with $\sum_{i=1}^{n+r} s_i = s$.

Conclusion. (i) The entire conclusions of Theorem 3.

(ii) The $\dfrac{\partial^t \varphi_i}{\partial \lambda_1^{t_1} \ldots \partial \lambda_r^{t_r}}$ exist and are continuous for $t_i \geq 0$, $\sum_{i=1}^{r} t_i = t$, and all $t \leq s$ with $x_0 \leq x \leq c$ and $|\lambda_k - \lambda_{k,0}| < a$, $k = 1, \ldots, r$.

7.2 The corollary to Theorem 3 also generalizes and we have the fact that the higher partials with respect to the λ's satisfy a linear differential equation with the same homogeneous part as the equation on $z_{ik} = \dfrac{\partial y_i}{\partial \lambda_k}$ and with zero initial conditions.

8. Higher derivatives with respect to x initial condition

8.1 Again the hypothesis of Theorem 6 augmented by the condition that $\partial^{s-1} f_i/\partial x^{s-1}$ be continuous is sufficient to prove the existence and continuity of

$$\frac{\partial^s \varphi_j}{\partial x^{*s}}.$$

8.2 We can combine all these results as follows.

Theorem 8. Hypothesis. (i) The entire hypothesis of Theorem 4.

(ii) Let $\dfrac{\partial^s f_i}{\partial y_1^{s_1} \ldots \partial y_n^{s_n} \partial \lambda_1^{s_{n+1}} \ldots \partial \lambda_r^{s_{n+r}}}$ exist and be jointly continuous in all variables for $s_j \geq 0$, $\sum_{j=0}^{n+r} s_j = s$.

Conclusion. (i) The entire conclusions of Theorem 4.

(ii) The partials $\dfrac{\partial^t \varphi_i}{\partial y_1^{*t_1} \ldots \partial y_n^{*t_n} \partial x^{*t_{n+1}} \partial \lambda_1^{t_{n+2}} \ldots \partial \lambda_r^{t_{n+r+1}}}$ exist and are continuous for all $t_j \geq 0$, $\sum_{j=1}^{n+r+1} t_j = t$, and all $t \leq s$, with $x_0 \leq x \leq c$, $|y_i^* - y_{i,0}| < b^+$, $|x^* - x_0| < b^+$, $|\lambda_k - \lambda_{k,0}| < a$.

9. Derivatives with respect to the independent variable

9.1 In previous sections of this chapter we examined the dependence of solution on initial conditions and parameters. We now wish to examine an analogous theorem for the dependence of solutions on the independent variable x. That is, given the differential equations

$$\frac{dy_i}{dx} = f_i(y_1, \ldots, y_n, x),$$

what conditions must we impose on the f_i in order that the solution

$$y_i = \varphi_i(x)$$

have continuous derivatives of any order. Once we have settled this problem we shall consider the question of *analyticity* of solutions in x and in parameters.

9.2 Suppose we fix an interval $x_0 \leq x \leq c$ and suppose as above that a solution is given whose locus is in a convex open region $\mathfrak{A}$. One can readily show that the existence and continuity of partials of orders,

$$\frac{\partial^s f_i}{\partial y_1^{s_1} \ldots \partial y_n^{s_n} \partial x^{s_0}}, \qquad s_0 + s_1 + \ldots + s_n = s$$

on a region including the locus associated with the solution, will imply the existence and continuity of the $s + 1$ derivatives of the solution $y_j = \varphi_j(x)$.

Let us then suppose, for instance, that $s = 1$; that is, $\frac{\partial f_i}{\partial y_j}$ and $\frac{\partial f_i}{\partial x}$ exist and are continuous on a region including the locus of points $\{\varphi_1(x), \ldots, \varphi_n(x), x\}$. But this means that

$$f_i(y_1, \ldots, y_n, x)$$

is a differentiable function of x having a continuous derivative when one substitutes differentiable functions of x for the $y_1, \ldots, y_n$. When one uses the given solution for this purpose, the result indicates that dy_i/dx is differentiable with a continuous derivative d^2y_i/dx^2. A straightforward induction readily yields the above stated result for derivatives of order $s + 1$.

10. Analyticity in solutions

10.1 We now turn to a very important theorem which can be stated roughly as follows: If the f_i are analytic in all variables then the solution is analytic in x.

When we say that a function f(x) is *analytic* we shall mean that f (x) can be expanded about some point $x = x_0$ in a Taylor series with a positive radius of convergence. Analyticity can of course be defined either in terms of real or complex variables. However, the ideas clustered about the concept of analyticity most naturally fall into the complex domain. Clearly, the complex case covers the real case also. So from now on in this chapter our variables will be assumed to be complex variables, and when we speak of a point x we mean $x = x_1 + ix_2$ as a point in two-dimensional real euclidean space. One remark on notation, $|x|$ is to be interpreted as the modulus of x.

We point out that our previous proofs, ostensibly for real variables, if interpreted properly, hold in the complex case.

10.2 Our theorem reads as follows:

Theorem 9. Hypothesis. Let $f_i(y_1, \ldots, y_n, x)$ be n complex valued functions of the $n + 1$ complex variables $y_1, \ldots, y_n$, x defined and analytic in a convex open region $\mathfrak{A}$ of $(n + 1)$-dimensional complex euclidean space (or $(2n + 2)$-dimensional real euclidean space).

Conclusion. For every complex point $(y_{1,0}, \ldots, y_{n,0}, x_0)$ of $\mathfrak{A}$ we can find a real $b^* > 0$ and n functions $y_1(x), \ldots, y_n(x)$ analytic in the circle $|x - x_0| \leq b^*$ such that

(i) $\dfrac{dy_i}{dx} = f_i(y_1(x), \ldots, y_n(x), x), \qquad i = 1, \ldots, n.$

(ii) $y_i(x_0) = y_{i,0}$.

(iii) The functions $y_i(x)$ are unique.

Comparing this theorem with Theorem 2 of Chapter 3 we note that in the hypothesis we have omitted the Lipschitz condition (which yielded us our uniqueness — (iii) of the conclusions). The Lipschitz condition is actually present, for analyticity of the f_i's implies the existence and continuity of the partials $\partial f_i/\partial y_j$ which in turn implies a Lipschitz condition.

10.3 The proof of the theorem proceeds according to the following plan. We define a closed region $\mathfrak{R}$ within $\mathfrak{A}$ and then a smaller region $\mathfrak{R}'$ such that the iterants have the locus of their solutions, that is, the set of points $\{y_1^{(k)}(x), \ldots, y_n^{(k)}(x), x\}$ in $\mathfrak{R}'$. We establish that the f_i are bounded in $\mathfrak{R}$ and satisfy a Lipschitz condition in $\mathfrak{R}'$. Then the usual Picard arguments show that the iterants converge uniformly to a solution and thus we can prove that the solution is analytic by showing that every iterant is analytic.

Since $\mathfrak{A}$ is open there exists a closed region $\mathfrak{R}$ around any point $y_{1,0}, \ldots, y_{n,0}, x_0$ in $\mathfrak{A}$, that is, there exists an $a > 0$ such that $|y_j - y_{j,0}| \leq a$, $|x - x_0| \leq a$ is in $\mathfrak{A}$. The f_i are continuous on $\mathfrak{R}$, and hence there exists a constant M such that $|f_i| \leq M$ on $\mathfrak{R}$. Let

$$b = \min\left(a, \frac{a}{2M}\right).$$

10.4 We shall show that in the region $\mathfrak{M}$ defined by the inequalities $|y_j - y_{j,0}| \leq \frac{a}{2}$, $|x - x_0| \leq a$, the f_i satisfy a Lipschitz condition. To do this we consider the values of the function $f_i(y_1, \ldots, y_n, x)$ for $|y_k - y_{k,0}| = a$, $|y_j - y_{j,0}| \leq \frac{a}{2}$, $j \neq k$ and $|x - x_0| \leq a$. The set of points $\{y_1, \ldots, y_n, x\}$ thus defined is a closed point set. Let M_{ik} be the maximum of $|f_i|$ on this set. Let $M = \max\limits_{i,k} M_{ik}$ and let $L = 4\,M/a$.

Now if $y_1', \ldots, y_n', x$ and $y_1'', \ldots, y_n'', x$ are any two points of $\mathfrak{M}$, then

$$f_i(y_1', \ldots, y_n', x) - f_i(y_1'', \ldots, y_n'', x) \tag{34}$$
$$= \sum_{k=0}^{n-1} [f_i(y_1'', \ldots, y_k'', y_{k+1}', \ldots, y_n', x) - f_i(y_1'', \ldots, y_{k+1}'', y_{k+2}', \ldots, y_n', x)].$$

In each term in brackets all but one variable is the same throughout and we can regard the other variables as parameters, that is

$$f_i(y_1'', \ldots, y_k'', y_{k+1}', \ldots, y_n', x) - f_i(y_1'', \ldots, y_{k+1}'', y_{k+2}', \ldots, y_n', x) = g_{ik}(y_k'') - g_{ik}(y_k')$$

where

$$g_{ik}(z) = f_i(y_1'', \ldots, y_{k-1}'', z, y_{k+1}', y_n', x).$$

Now, regarded as a function of z, $g_{ik}(z)$ is analytic for $|z - y_{k,0}| \leq a$. Let C denote the circle in the z-plane given by $|z - y_{k,0}| = a$. The points y_k'' and y_k' are such that $|y_k'' - y_{k,0}| \leq \frac{a}{2}$ and $|y_k' - y_{k,0}| \leq \frac{a}{2}$.

If we apply Cauchy's Integral Formula,

$$g_{lk}(y_k'') - g_{lk}(y_k') = \frac{1}{2\pi i}\int_C \frac{g_{lk}(\zeta)}{\zeta - y_k''}\,d\zeta - \frac{1}{2\pi i}\int_C \frac{g_{lk}(\zeta)}{\zeta - y_k'}\,d\zeta$$
$$= \frac{1}{2\pi i}(y_k'' - y_k')\int_C \frac{g_{lk}(\zeta)}{(\zeta - y_k'')(\zeta - y_k')}\,d\zeta.$$

From the definition of C and limitations on y_k' and y_k'', one can show that $|\zeta - y_k''| \geq \frac{a}{2}$, $|\zeta - y_k'| \geq \frac{a}{2}$ for every ζ on C. Furthermore, our definition of M_{lk} insures that $|g_{lk}(\zeta)| \leq M_{lk} \leq M$ for ζ on C. Consequently

$$|g_{lk}(y_k'') - g_{lk}(y_k')| \leq \frac{1}{2\pi}|y_k'' - y_k'|\frac{M}{(a/2)^2}(2\pi a) = L\,|y_k'' - y_k'|.$$

This result and Equation (34) show that

$$|f_l(y_1'', \ldots, y_n'', x) - f_l(y_1', \ldots, y_n', x)| \leq L\sum_{j=1}^{n}|y_j'' - y_j'|.$$

10.5 Let

$$y_1^{(0)}(x), \ldots, y_n^{(0)}(x) \tag{35}$$

be any set of n analytic functions defined on $|x - x_0| \leq b$ and such that $|y_j^{(0)}(x) - y_{j,0}| \leq \frac{a}{2}$ for this x circle. Furthermore let these functions pass through the point $y_{1,0}, \ldots, y_{n,0}, x_0$ in $\mathfrak{A}$; that is, $y_i^{(0)}(x_0) = y_{i,0}$. Define the Picard iterants

$$y_i^{(k+1)}(x) = y_{i,0} + \int_{x_0}^{x} f_i(y_1^{(k)}(t), \ldots, y_n^{(k)}(t), t)\,dt, \quad k = 0, 1, \ldots . \tag{36}$$

One can show inductively that they exist for $|x - x_0| \leq b$. The integral can be taken along any rectifiable path between x_0 and x which remains interior to the circle $|x - x_0| \leq b$. The integral is independent of path because of the analyticity, and hence (because we desire inequalities) we shall take the shortest path, namely a straight line. Furthermore, in this x range they satisfy the conclusion

$$|y_j^{(k)}(x) - y_{j,0}| \leq M\,|x - x_0| \leq Mb \leq M\frac{a}{2M} = \frac{a}{2}. \tag{37}$$

Equation (36) can also be used to establish inductively that the iterants are all analytic for $|x - x_0| \leq b$. (The proof is formally the same as that of Lemma 2 of Chapter 4 above.)

Thus the iterants have a locus $\{y_1^{(k)}(x), \ldots, y_n^{(k)}(x), x\}$ which lies in $\mathfrak{M}$ for $|x - x_0| \leq b$. Hence we can apply the usual arguments for Picard iterants and show that these iterants converge uniformly for $|x - x_0| \leq b$ to a solution. (Cf. Theorem 2 of Chapter 4 above.) Since the iterants are analytic and converge uniformly the limit is analytic in $|x - x_0| \leq b^*$ for any $b^* > 0$ and less than b. This completes the proof of the theorem.

11. Analyticity in parameters

11.1 Because of the great practical importance of parameters in the theory of differential equations, we state the following theorem.

Theorem 10. Hypothesis. Let $f_i(y_1, \ldots, y_n, x, \lambda_1, \ldots, \lambda_r)$ be n complexed valued functions of the $n + 1 + r$ complex variables $y_1, \ldots, y_n, x, \lambda_1, \ldots, \lambda_r$ defined and analytic in a convex open region $\mathfrak{A}$ of $(n + 1 + r)$-dimensional complex euclidean space.

Conclusion. For every complex point $(y_{1,0}, \ldots, y_{n,0}, x_0, \lambda_{1,0}, \ldots, \lambda_{r,0})$ of $\mathfrak{A}$ we can find a real $b^* > 0$ and n functions $y_1(x, \lambda_1, \ldots, \lambda_r), \ldots, y_n(x, \lambda_1, \ldots, \lambda_r)$ analytic in the circle $|x - x_0| \leq b^*$ and with $\lambda_{i,0} - \lambda_i | \leq a$ for some $a > 0$ such that

(i) $\dfrac{dy_i}{dx} \equiv f_i(y_1, \ldots, y_n, x, \lambda_1, \ldots, \lambda_r), \qquad i = 1, \ldots, n.$

(ii) $y_{j,0} = y_j(x_0, \lambda_1, \ldots, \lambda_r)$.

(iii) The functions $y_j(x)$ are unique.

The proof of this theorem parallels that of Theorem 9.

12. α parameter theory

12.1 In this section we wish to give a practical example of parameter theory in the theory of ordinary differential equations. Suppose we have a differential equation

$$\frac{dy}{dx} = f(y, x, \alpha)$$

which depends on a parameter α. We shall assume f is analytic in the neighborhood of a point (y_0, x_0, α_0). The solution of this equation is

$$y = \varphi(x, \alpha)$$

which is analytic in some circle $|x - x_0| \leq b$ with $|\alpha - \alpha_0| \leq a$, and at the initial point

$$y_0 = \varphi(x_0, \alpha).$$

12.2 Such a situation may arise in the following fashion. Suppose we wish to solve the differential equation

$$\frac{dy}{dx} = f(y, x, \alpha_0)$$

on a machine. Now due to various imperfections in the machine, the equation as actually realized on the machine is

$$\frac{dy}{dx} = f(y, x, \alpha).$$

Hence the machine solution is $\varphi(x, \alpha)$ instead of the actual solution $\varphi(x, \alpha_0)$. We wish, therefore, to determine the effects of the machine error, that is, we wish to determine $\varphi(x, \alpha) - \varphi(x, \alpha_0)$. The parameter theory we shall develop here based on the fundamental theorems of this chapter enable us to do this without the customary — not necessarily valid — "linearization."

12.3 The machine solution is $\varphi(x, \alpha)$. Our theory tells us that $\varphi(x, \alpha)$ is analytic in α and hence may be expanded in a Taylor series about the point $\alpha = \alpha_0$. This yields

$$\varphi(x, \alpha) = \sum_{k=0}^{\infty} \frac{\partial^k}{\partial \alpha^k} \varphi(x, \alpha_0) \frac{(\alpha - \alpha_0)^k}{k!}.$$

Our problem is thus to determine the $\frac{\partial^k}{\partial \alpha^k} \varphi(x, \alpha_0)$ functions. Now let

$$z(x, \alpha) = \frac{\partial}{\partial \alpha} \varphi(x, \alpha).$$

We have seen that z satisfies the linear differential equation

$$\frac{dz}{dx} = \frac{\partial f}{\partial y} z + \frac{\partial f}{\partial \alpha}$$

with the initial condition $z(x_0, \alpha) = 0$. Hence at $\alpha = \alpha_0$.

$$z(x, \alpha_0) = \frac{\partial}{\partial \alpha} \varphi(x, \alpha_0)$$

and we have the differential equation

$$\frac{dz}{dx} = \left(\frac{\partial f}{\partial y}\right)_{\alpha_0} z + \left(\frac{\partial f}{\partial \alpha}\right)_{\alpha_0}. \tag{38}$$

Clearly the quantities on the right are known since they are derived from the given function f. The solution of Equation (38) is obtained with the aid of the integrating factor

$$E(x) = \exp\left[\int_{x_0}^{x} \frac{\partial f}{\partial y}\, dt\right]$$

to be

$$z(x, \alpha_0) = E(x) \int_{x_0}^{x} \frac{\partial}{\partial \alpha} f(t, \alpha_0)\, E^{-1}(t)\, dt\ . \tag{39}$$

Hence we have determined $\frac{\partial}{\partial \alpha} \varphi(x, \alpha_0)$.

Now we know that the *second* partial

$$w = \frac{\partial^2}{\partial \alpha^2} \varphi(x, \alpha_0)$$

satisfies a linear differential equation with the *same* homogeneous part as the equation on x. Explicitly

$$\frac{dw}{dx} = \left(\frac{\partial f}{\partial y}\right)_{\alpha_0} w + \left[\left(\frac{\partial^2 f}{\partial y^2}\right)_{\alpha_0} z^2(x, \alpha_0) + \left(\frac{\partial^2 f}{\partial \alpha^2}\right)_{\alpha_0}\right]. \tag{40}$$

The expression in the brackets is known since $\partial^2 f/\partial y^2$ and $\partial^2 f/\partial \alpha^2$ depend on the given function f, and z has been determined from Equation (39). Equation (40) is immediately solvable and we have determined the second partials

$$w(x, \alpha_0) = \frac{\partial^2}{\partial \alpha^2} \varphi(x, \alpha_0).$$

Proceeding in this way, all the terms of the Taylor series expansion can be determined.

12.4 In the case of more than one dependent variable, the resulting linear differential system may not be solvable by quadratures; however, we note that all the partials $\partial^s \varphi/\partial \alpha^s$ depend on the solutions of only *one linear* differential equation system.

CHAPTER 6

Linear Differential Equations

1. Introduction

1.1 In our previous discussions we have considered systems of differential equations in the form

$$y_j' = f_j(y_1, \ldots, y_n, x), \qquad j = 1, \ldots, n. \tag{1}$$

We have shown that as more and more properties were assumed for the f_j functions, more conclusions could be drawn concerning the solutions. However, in every case up to now, our existence result could be applied only to a neighborhood of the initial point x_0, which appears to be smaller than the region one would associate with our hypotheses on the f_j. The example $y' = y^2$ (cf. Chapter 1, Section 1.8) shows that the restriction of the solution to a neighborhood is essential even when f is a polynomial in y. Now there is only one type of equation in which y appears in a simpler way, that is, an equation linear in y. Here at last the restriction "in the small" is not necessary. As we shall show in the present chapter, the system

$$\dot{y}_i = \sum_{j=1}^{n} p_{ij}(x)y_j + q_i(x) \text{ [1]} \tag{2}$$

will have a unique solution $y_1(x), \ldots, y_n(x)$, with specified values at x_0, defined on any closed interval containing x_0 for which $p_{ij}(x)$ and $q_i(x)$ are continuous for every i and j.

In order to accomplish this, however, we must first discuss the case where $q_i(x) \equiv 0$, that is, the system

$$\dot{y}_i = \sum_{j=1}^{n} p_{ij}(x)y_j, \quad i = 1, \ldots, n \tag{3}$$

[1] We are now using another conventional notation, viz.:

$$\dot{y} \equiv \frac{dy}{dx}$$

which is termed homogeneous. A solution of Equation (3) is an n-tuple of functions $\{y_1(x), \ldots, y_n(x)\}$. The system of Equation (3) obviously has the characteristic that if $\{z_1(x), \ldots, z_n(x)\}$ is another solution then $\{c_1y_1(x) + c_2z_1(x), \ldots, c_1y_n(x) + c_2z_n(x)\}$ is also a solution. If the first n-tuple is denoted by Y and the second by Z, then the third can be denoted by $c_1Y + c_2Z$. Thus the set of solutions of Equation (3) defined on any closed interval for which the p_{ij} are continuous constitute a linear vector space. We shall show that in this case the linear vector space is n dimensional. This is of course equivalent to the statement that every solution is a linear combination of n specified solutions, which are linearly independent.

The process of choosing n linearly independent solutions corresponds to choosing n linearly independent directions in the vector space of solutions, that is, choosing a coordinate system. We shall show that n linearly independent solutions

$$Y_j = \{y_{1j}(x), \ldots, y_{nj}(x)\}, \qquad j = 1, \ldots, n \tag{4}$$

can be chosen so that for every x_0 in the specified interval the determinant of the matrix

$$\| y_{ij}(x_0) \| \tag{5}$$

is unequal to zero. But this immediately implies that given any system of n initial values $\{y_{1,0}, \ldots, y_{n,0}\}$, there is a linear combination $c_1Y_1 + \ldots + c_nY_n$ which assumes these values at x_0. For the determinant of the equations

$$y_{i,0} = \sum_j c_j\, y_{ij}(x_0) \tag{6}$$

on $c_1, \ldots, c_n$ is not zero and thus these equations have a unique solution which furnishes the required linear combination $c_1Y_1 + \ldots + c_nY_n$ which is the solution of Equation (3) with the specified initial values.

Thus the problem of solving Equation (3) consists in obtaining n linearly independent solutions with the specified property on the determinant of values. This will be done in Section 2 below.

1.2 The corresponding problem for the system of Equation (2) can now be solved if we know merely one solution of Equation (2),

$\{v_1(x), \ldots, v_n(x)\}$ with any initial conditions. For let $\{y_{1,0}, \ldots, y_{n,0}\}$ again be a set of initial values corresponding to $x = x_0$. We know from the above that there is a solution of Equation (3) $\{z_1(x), \ldots, z_n(x)\}$ with initial values $z_j(x_0) = y_{j,0} - v_j(x_0)$. Let

$$y_j = z_j + v_j. \tag{7}$$

Clearly the y_j will have the specified values at x_0. Furthermore, if we substitute $y_j = z_j + v_j$ in Equation (2), then

$$\begin{aligned} \dot{y}_j = \dot{z}_j + \dot{v}_j &= \sum_j p_{ij} z_j + (\sum_j p_{ij} v_j + q_i) = \sum_j p_{ij}(z_j + v_j) + q_i \\ &= \sum_j p_{ij}\, y_j + q_i. \end{aligned}$$

Thus when the homogeneous problem has been solved, the non-homogeneous problem is merely a matter of finding one solution. In addition, as we shall see, the solution of the homogeneous system will permit us to find one solution of the non-homogeneous system by simple quadratures.

1.3 This is then the situation for a real variable x. The basis of this discussion is a certain formal investigation of the determinant of values which appears in Equation (5). The formal investigation can also be applied in the case of a complex variable. Suitably joined with the Jordan theorem for matrices, the multi-valued character of the solutions can then be specified.

1.4 We also consider two special cases of great interest. One is the case of a linear differential equation of the nth order on only one dependent variable, the other is the important practical case of linear differential equations with constant coefficients.

2. Existence and uniqueness

2.1 As we have remarked in the previous section, linear differential equations are the only practical class of differential equations for which we can obtain theorems *in the large*. We shall first prove the theorem for the real case and later indicate the complex case analog.

Theorem 1. Hypothesis. Let

$$\dot{y}_i = \sum_{j=1}^{n} p_{ij}(x) y_j, \qquad i = 1, \ldots, n \tag{8}$$

be a system of linear differential equations where the $p_{ij}(x)$ are defined and continuous on some closed interval I: $\{a \leq x \leq b\}$ of the x-axis.

Conclusion. Let x_0 be any point of I. Let $y_{1,0}, \ldots, y_{n,0}$ be any set of n real numbers. Then there exists a unique solution of Equation (8),

$$\Phi_1(x), \ldots, \Phi_n(x)$$

defined over [a, b] and satisfying the boundary conditions

$$\Phi_i(x_0) = y_{i,0}, \qquad i = 1, \ldots, n.$$

Proof. Consider

$$\dot{y}_i = \sum_{j=1}^{n} p_{ij}\, y_j \equiv f_i(y_1, \ldots, y_n, x). \tag{9}$$

Since the $p_{ij}(x)$ are continuous on I, they are bounded. Let

$$| p_{ij}(x) | \leq L_{ij}$$

and let

$$L = \max_{i,j} L_{ij}.$$

2.2 We shall now prove that the f_j functions satisfy a Lipschitz condition on I. A Lipschitz condition was not postulated in the hypothesis, since in the linear case we can establish its existence. This is indeed the origin of the Lipschitz condition postulated in the general case (cf. Theorem 2 of Chapter 3 and Theorem 2 of Chapter 4).

Let $y_1^*, \ldots, y_n^*$ and $y_1^+, \ldots, y_n^+$ be any two sets of n real numbers. Then

$$\begin{aligned} &| f_i(y_1^*, \ldots, y_n^*, x) - f_i(y_1^+, \ldots, y_n^+, x) | \\ &\quad = | p_{i1}(y_1^* - y_1^+) + p_{i2}(y_2^* - y_2^+) + \cdots + p_{in}(y_n^* - y_n^+) | \\ &\quad \leq | p_{i1} || y_1^* - y_1^+ | + | p_{i2} || y_2^* - y_2^+ | + \cdots + | p_{in} || y_n^* - y_n^+ | \\ &\quad \leq L \sum_{j=1}^{n} | y_j^* - y_j^+ |, \end{aligned}$$

which is our Lipschitz condition.

2.3 Let x′ be any point in I and consider the following set of initial conditions:

$$(y_{1,0}, \ldots, y_{n,0}, x') = (\delta_{1k}, \delta_{2k}, \ldots, \delta_{nk}, x')$$

for some k, $k = 1, \ldots, n$. (δ_{ij} is the Kronecker delta.) So we have

$$y_i(x') = \delta_{ik}.$$

Consider the region $\mathfrak{A}$ in $(n + 1)$-dimensional euclidean space for which

$$|y_i - \delta_{ik}| \leq 1, \quad |x' - x| \leq 1 \quad \text{and} \quad x \varepsilon I.$$

Observe that $|y_i - \delta_{ii}| \leq 1$ implies $0 \leq y_i \leq 2$ and $|y_i - \delta_{ik}| \leq 1$ for $k \neq i$ implies $-1 \leq y_i \leq 1$. Now

$$|f_i| = |\sum_j p_{ij} y_j| \leq \sum_j |p_{ij}|\,|y_j| \leq \sum_j L\,|y_j| = L \sum_j |y_j|$$

and on the region $\mathfrak{A}$,

$$|f_i| \leq L \sum_{j=1}^{n} |y_j| \leq L + L + \ldots + \underbrace{2L}_{k^{th}\text{ term}} + \ldots + L = (n+1)L.$$

2.4 Define

$$b^* = \min\left(\frac{1}{L(n+1)}, |x' - a|, |b - x'|\right).$$

(The last two terms $|x' - a|$ and $|b - x'|$ insure us that our x interval $|x - x'| \leq b^*$ remains in I.)

We can, as earlier, construct the Picard iterants. The mth iterant is given by

$$y_i^{(m)}(x) = \delta_{ik} + \int_{x'}^{x} f_i^{(m-1)} dt, \qquad i = 1, \ldots, n$$

and if $|x - x'| \leq b^*$,

$$|y_i^{(m)}(x) - \delta_{ik}| \leq \left| \int_{x'}^{x} |f_i^{(m-1)}|\, dt \right| \leq L(n + 1)\, b^* \leq 1.$$

So if the first iterant is in the interval $|x - x'| \leq b^*$, then all the Picard iterants are. Further, by the Lipschitz condition, the iterants converge uniformly in $|x - x'| \leq b^*$ to a unique solution of the given differential equation with the initial conditions $y_i(x') = \delta_{ik}$.

2.5 Two immediate corollaries of the above result are:

Corollary 1. If the interval $I' : \left\{x' \leq x \leq x' + \frac{1}{L(n+1)}\right\}$ is in I, then the Picard existence theorem holds on I′ with the initial conditions $y_i(x') = \delta_{ik}$.

Corollary 2. If $x' \leq b < x' + \frac{1}{L(n+1)}$, then the Picard existence theorem holds on $x' \leq x \leq b$ with the initial conditions $y_i(x') = \delta_{ik}$.

Similar corollaries hold for $x < x'$.

2.6 So far we have the usual existence and uniqueness theorem *in the small.* That is, we have a solution valid in a neighborhood of x'. We shall now show how the linearity of the equations enables us to extend this solution to the whole interval, $a \leq x \leq b$.

Let

$$\lambda = \frac{1}{L(n+1)}.$$

Let $x_1 = x_0 + \lambda$ provided $x_0 + \lambda \varepsilon I$, otherwise let $x_1 = b$. If $x_1 \varepsilon I$, let $x_2 = x_1 + \lambda$ provided $x_1 + \lambda \varepsilon I$, otherwise let $x_2 = b$. We can continue this process m times until we finally get $x_m = b$.

Now let x_0 (the given initial point) equal the x' of the previous discussion. At x_0 we have the solutions

$$\varphi_{1k}, \ldots, \varphi_{nk}$$

which at $x = x_0$ have the initial values $\varphi_{ik}(x_0) = \delta_{ik}$. By Corollary 1 the $\varphi_{ik}(x)$, $i = 1, \ldots, n$, any k, are all uniquely defined throughout $x_0 \leq x \leq x_1$ (or by Corollary 2 if $x_1 = b$).

Since we are dealing with a *linear* system, any linear combination (with constant coefficients) of the $\varphi_{ik}(x)$ functions is also a solution of Equation (9). In particular, if we use the initial conditions $y_{i,0}$ as multiplicative constants,

$$\sum_{k=1}^{n} y_{k,0}\, \varphi_{ik}(x), \qquad i = 1, \ldots, n$$

is a solution defined throughout $x_0 \leq x \leq x_1$. Furthermore, this solution has the initial condition $y_{1,0}, \ldots, y_{n,0}$ at x_0 because if we let

$$\varphi_i(x) = \sum_{k=1}^{n} y_{k,0}\, \varphi_{ik}(x), \qquad i = 1, \ldots, n \tag{10}$$

then

$$\varphi_i(x_0) = \sum_k y_{k,0}\, \varphi_{ik}(x_0) = \sum_k y_{k,0}\, \delta_{ik} = y_{i,0}.$$

By our Lipschitz condition, these solutions $\varphi_i(x)$ through $y_{1,0}, \ldots,$ $y_{n,0}$, x_0 are unique on $x_0 \leq x \leq x_1$.

Now if $x_1 = b$, the theorem is proved for $x_0 \leq x \leq b$. If $x_1 = x_0 + \lambda < b$, we proceed as follows. The solution $\varphi_i(x)$ is defined for the $[x_0, x_1]$ interval. Hence $\varphi_1, \ldots, \varphi_n$ assume some set of values at $x = x_1$, say $y_{1,1}, \ldots, y_{n,1}$.

At x_1 we have a similar set of solutions to those we had at x_0 (that is, we let the x' of the corollaries now be x_1). Let these solutions be

$$\psi_{1k}(x), \ldots, \psi_{nk}(x)$$

which at $x = x_1$ have the initial values $\psi_{ik}(x_1) = \delta_{ik}$. The $\psi_{ik}(x)$ are uniquely determined throughout $x_1 \leq x \leq x_2$. Now any linear combination of the ψ_{ik} functions is also a solution. In particular

$$\psi_i(x) = \sum_{k=1}^{n} y_{k,1}\, \psi_{ik}(x), \qquad i = 1, \ldots, n$$

is such a set of solutions of Equation (9) defined throughout $x_1 \leq x \leq x_2$ and such that at $x = x_1$,

$$\psi_i(x_1) = y_{i,1}.$$

Hence we have a set of unique solutions $\varphi_i(x)$ defined on $[x_0, x_1]$ and a set of unique solutions $\psi_i(x)$ defined on $[x_1, x_2]$ which at the common point $x = x_1$ have the same value, that is,

$$\varphi_i(x_1) = y_{i,1} = \psi_i(x_1).$$

Also, since $\varphi_i(x)$ and $\psi_i(x)$ both satisfy Equation (9) at $x = x_1$,

$$\dot{\varphi}_i(x_1) = \dot{\psi}_i(x_1).$$

Therefore, $\chi_i(x)$, $i = 1, \ldots, n$ where

$$\begin{aligned} \chi_i(x) &= \varphi_i(x) \quad \text{for} \quad x_0 \leq x \leq x_1 \\ \chi_i(x) &= \psi_i(x) \quad \text{for} \quad x_1 \leq x \leq x_2 \end{aligned}$$

is the unique solution through $y_{1,0}, \ldots, y_{n,0}, x_0$ defined throughout $x_0 \leq x \leq x_2$.

If $x_2 = b$, then the theorem is proved for $x_0 \leq x \leq b$. If not, we can continue the process a finite number of times, m, continuing the solution at each step until we finally obtain a unique solution, $\Omega_i(x)$,

$$\Omega_i(x) = \varphi_i(x) \quad \text{on} \quad [x_0, x_1]$$
$$\Omega_i(x) = \psi_i(x) \quad \text{on} \quad [x_1, x_2]$$
$$\cdots\cdots\cdots\cdots\cdots$$
$$\Omega_i(x) = \omega_i(x) \quad \text{on} \quad [x_{m-1}, b]$$

through $y_{1,0}, \ldots, y_{n,0}, x_0$ defined throughout $x_0 \leq x \leq b$.

Similarly we can define a solution $\Lambda_i(x)$ defined throughout $a \leq x \leq x_0$ which is the unique solution on this interval through the point $y_{1,0}, \ldots, y_{n,0}, x_0$. So,

$$\Lambda_i(x_0) = y_{i,0} = \Omega_i(x_0)$$

and hence $\Phi_i(x)$,

$$\Phi_i(x) = \Omega_i(x) \quad \text{for} \quad x_0 \leq x \leq b$$
$$\Phi_i(x) = \Lambda_i(x) \quad \text{for} \quad a \leq x \leq x_0$$

is defined for $a \leq x \leq b$ and it is the unique solution through $y_{1,0}, \ldots, y_{n,0}, x_0$ on this interval.

2.7 The corresponding theorem for the complex case is:

Theorem 2. Hypothesis. Let C be a simple rectifiable arc. Let

$$\dot{y}_i = \sum_{j=1}^{n} p_{ij}(x)\, y_j, \qquad i = 1, \ldots, n \tag{11}$$

where the $p_{ij}(x)$ are analytic functions of x on C.

Conclusion. Let x_0 be any point on C. Let $y_{1,0}, \ldots, y_{n,0}$, be any set of n complex numbers. Then there exists a unique solution of Equation (11) defined and analytic on C which at $x = x_0$ assumes the given initial values, $y_{1,0}, \ldots, y_{n,0}$.

The proof of this theorem is analogous to the proof in the real case (Theorem 1). Note that in the complex case the proof requires analytic continuation.

3. Wronskian theory

3.1 The determinant of Equation (5) consisting of the values of a set of n solutions of a system of linear differential equations plays a fundamental role in our theory. This determinant is called the Wronskian. It will be suitably defined and discussed in the present section.

3.2 For the real case:

Theorem 3. Hypothesis. Let

$$\dot{y}_i = \sum_{j=1}^{n} p_{ij}(x) y_j, \qquad i = 1, \ldots, n \tag{12}$$

where the $p_{ij}(x)$ are defined and continuous on the x interval [a, b].

Conclusion. There exist n solutions

$$\begin{array}{c} y_{11}(x), \ldots, y_{n1}(x) \\ y_{12}(x), \ldots, y_{n2}(x) \\ \cdots\cdots\cdots \\ y_{1n}(x), \ldots, y_{nn}(x) \end{array} \tag{13}$$

of the given system such that

$$W(x) = \begin{vmatrix} y_{11}(x) & \ldots & y_{n1}(x) \\ y_{12}(x) & \ldots & y_{n2}(x) \\ \cdot & \cdots & \cdot \\ y_{1n}(x) & \ldots & y_{nn}(x) \end{vmatrix}$$

is never zero on $a \leq x \leq b$ and W(x) satisfies the differential equation

$$\dot{W} - \left[\sum_{i=1}^{n} p_{ii}(x)\right] W = 0.$$

The determinant W(x) is called the *Wronskian* of the solutions, Equation (13), or the Wronskian of the system of linear differential equations, Equation (12).

3.3 The function W(x) has a continuous derivative on I: [a, b] since the $y_{ij}(x)$ functions have this property and W is a polynomial function of the y_{ij}. Consider, then, the derivative $\dot{W}(x)$ of W(x). It is:

$$\dot{W}(x) = \sum_{i=1}^{n} \begin{vmatrix} y_{11} & \cdots & y_{i-1,1} & \dot{y}_{i1} & y_{i+1,1} & \cdots & y_{n1} \\ y_{12} & \cdots & y_{i-1,2} & \dot{y}_{i2} & y_{i+1,2} & \cdots & y_{n2} \\ \cdot & \cdot & \cdot & \cdot & \cdot & \cdot & \cdot \\ y_{1n} & \cdots & y_{i-1,n} & \dot{y}_{in} & y_{i+1,n} & \cdots & y_{nn} \end{vmatrix} = \sum_{i=1}^{n} W_i.$$

Now, by Equation (12) we may replace $\dot{y}_{ik}$, i, k = 1, . . ., n by

$$\sum_{j=1}^{n} p_{ij}(x)\, y_{jk}.$$

Hence

$$W_i(x) = \begin{vmatrix} y_{11} \cdots \sum_{j=1}^{n} p_{ij}\, y_{j1} \cdots y_{n1} \\ y_{12} \cdots \sum_{j=1}^{n} p_{ij}\, y_{j2} \cdots y_{n2} \\ \cdot \quad \cdot \quad \cdot \quad \cdot \quad \cdot \quad \cdot \quad \cdot \\ y_{1n} \cdots \sum_{j=1}^{n} p_{ij}\, y_{jn} \cdots y_{nn} \end{vmatrix}$$

Now each $\sum_{\substack{j=1 \\ j \neq i}}^{n} p_{ij} y_{jk}$ is a linear combination of the remaining columns and hence can be eliminated by the usual rules for manipulating determinants. Therefore

$$W_i = \begin{vmatrix} y_{11} \cdots p_{ii}y_{i1} \cdots y_{n1} \\ y_{12} \cdots p_{ii}y_{i2} \cdots y_{n2} \\ \cdot \quad \cdot \quad \cdot \quad \cdot \quad \cdot \quad \cdot \\ y_{1n} \cdots p_{ii}y_{in} \cdots y_{nn} \end{vmatrix} = p_{ii}\, W.$$

Hence

$$\dot{W}(x) = \sum_{i=1}^{n} W_i(x) = W(x) \sum_{i=1}^{n} p_{ii}(x) = t(x)\, W(x)$$

where $t(x)$ is the trace of the matrix of the determinant $W(x)$. We therefore have

$$\dot{W} = t(x)\, W, \tag{14}$$

which completes the proof of the theorem.

With the aid of the integrating factor

$$e^{\int_{x_0}^{x} t(\xi)\, d\xi}$$

we can integrate Equation (14) yielding

$$W(x) = C_0\, e^{\int_{x_0}^{x} t(\xi)\, d\xi}. \tag{15}$$

3.4 Consider now the n solutions $\varphi_{jk}(x)$, $k = 1, \ldots, n$ of the given system, Equation (12), for which

$$\varphi_{jk}(x_0) = \delta_{jk}, \qquad j, k = 1, \ldots, n$$

and let $x_0 = a$. From Theorem 1, these φ_{jk} functions with the above initial conditions are uniquely defined on I. For this particular system we have from Equation (15)

$$W(a) = C_0\, e^{\int_a^a t(\xi)\, d\xi} = C_0.$$

Or in more expanded notation,

$$C_0 = W(a) = \begin{vmatrix} \delta_{11} & \delta_{21} & \dots & \delta_{n1} \\ \delta_{12} & \delta_{22} & \dots & \delta_{n2} \\ \cdot & \cdot & \cdot & \cdot \\ \delta_{1n} & \delta_{2n} & \dots & \delta_{nn} \end{vmatrix} = 1.$$

Therefore

$$W(x) = e^{\int_{x_0}^{x} t(\xi)\, d\xi}$$

and this can never be zero since $\int_{x_0}^{x} t(\xi)\, d\xi$ is a finite quantity and the exponential is never zero.

3.5 The complex analog of Theorem 3 is:

Theorem 4. Hypothesis. Let

$$\dot{y}_i = \sum_{j=1}^{n} p_{ij}(x)\, y_j, \qquad i = 1, \dots, n$$

where the $p_{ij}(x)$ are defined and analytic on a simple rectifiable arc C.

Conclusion. There exist n solutions

$$\begin{matrix} y_{11}(x), \dots, y_{n1}(x) \\ y_{12}(x), \dots, y_{n2}(x) \\ \cdot \quad \cdot \quad \cdot \quad \cdot \\ y_{1n}(x), \dots, y_{nn}(x) \end{matrix}$$

of the given system such that

$$W(x) = \begin{vmatrix} y_{11}(x) & \dots & y_{n1}(x) \\ \cdot & \cdot & \cdot \\ y_{1n}(x) & \dots & y_{nn}(x) \end{vmatrix}$$

is never zero for x on C.

3.6 Suppose we have

$$\dot{y}_i = \sum_{j=1}^{n} p_{ij}(x)\, y_j, \qquad i = 1, \dots, n$$

where the $p_{ij}(x)$ are continuous on $a \leq x \leq b$. If

$$\begin{matrix} y_{11}(x), \ldots, y_{n1}(x) \\ y_{12}(x), \ldots, y_{n2}(x) \\ \cdots\cdots\cdots \\ y_{1n}(x), \ldots, y_{nn}(x) \end{matrix}$$

are n solutions of the given system on $a \leq x \leq b$ such that $W(x) \neq 0$ on $a \leq x \leq b$ then this system of solutions will be called a basic system of solutions or *a fundamental system of solutions.*

Similarly for the complex case.

3.7 We shall now show that all basic systems of solutions are equivalent.

Theorem 5. Hypothesis. Let

$$\dot{y}_i = \sum_{j=1}^{n} p_{ij}(x)\, y_j, \qquad i = 1, \ldots, n \tag{16}$$

where the $p_{ij}(x)$ are continuous on I: [a, b]. Let

$$\begin{matrix} z_{11}, \ldots, z_{n1} \\ z_{12}, \ldots, z_{n2} \\ \cdots\cdots \\ z_{1n}, \ldots, z_{nn} \end{matrix} \tag{17}$$

be any fundamental system of solutions of Equation (16) on I.

Conclusion. Every other solution of Equation (16) on I is a linear combination with constant coefficients of the given basic system of solutions.

The proof of this theorem can be found in Section 1.1 in connection with the discussion of Equations (5) and (6). A similar theorem holds in the complex case.

3.8 Since any fundamental system is a linear combination of any other fundamental system, we have:

Corollary. Let $\varphi_{ik}(x)$ and $\psi_{ik}(x)$ be two fundamental systems of solutions of the linear differential equations

$$\dot{y}_i = \sum_{j=1}^{n} p_{ij}(x)\, y_j, \qquad i = 1, \ldots, n. \tag{18}$$

Then the Wronskians $W_\varphi(x)$ and $W_\psi(x)$ of these two fundamental systems differ by a non-zero multiplicative constant, that is,

$$W_\varphi = K\, W_\psi, \quad K \neq 0. \tag{19}$$

4. Variation of parameters and the Green's function

4.1 In previous theorems we have dealt with *homogeneous* linear differential equations. We now wish to consider the non-homogeneous case, namely

$$\dot{y}_i = \sum_{j=1}^{n} p_{ij}(x)\, y_j + q_i(x) \quad , \quad i = 1, \ldots, n. \tag{20}$$

An extension of the familiar method of "variation of parameters" enables us rigorously to solve this system and also leads to the convenient concept of the one-sided Green's function matrix.

Suppose $p_{ij}(x)$, $q_i(x)$ are continuous in some interval I: [a, b] of the x-axis and we wish to solve Equation (20) with the initial conditions

$$y_i(x_0) = y_{i,0} \quad \text{for} \quad x_0 \varepsilon I.$$

Suppose further we have functions $\varphi_i(x)$ such that $\varphi_i(x_0) = 0$ and

$$\dot{\varphi}_i = \sum_{j=1}^{n} p_{ij}(x)\, \varphi_j + q_i, \qquad i = 1, \ldots, n. \tag{21}$$

Then if y_{jk} is a fundamental system of solutions of the homogeneous equation corresponding to Equation (20) with initial conditions

$$y_{jk}(x_0) = \delta_{jk},$$

then

$$\psi_i(x) = \sum_{k=1}^{n} y_{k,0}\, y_{ik}(x) + \varphi_i(x), \qquad i = 1, \ldots, n$$

is a solution of Equation (20) with the desired initial conditions — as we have seen in Section 1.1 above.

Hence, if we have *a* solution $\varphi_i(x)$ of the non-homogeneous equation (called the "particular solution" in the elementary theory) with the initial conditions $\varphi_i(x_0) = 0$, $i = 1, \ldots, n$, then the solution of the differential equation, Equation (20), with arbitrary initial conditions, is $\varphi_i(x)$ plus a linear combination of a fundamental system. This solution is unique by our Picard existence and uniqueness theorem.

4.2 Our theorem for the solution of non-homogeneous equations can be stated as follows:

Theorem 6. Hypothesis. Let

$$\dot{y}_i = \sum_{j=1}^{n} p_{ij}(x)\, y_j + q_i(x), \qquad i = 1, \ldots, n \tag{22}$$

be a system of linear differential equations where the $p_{ij}(x)$ and $q_i(x)$ are continuous on I: [a, b].

Conclusion. The solution $\varphi_i(x)$ of Equation (22) with the initial condition $\varphi_i(x_0) = 0$, $x_0 \in I$ can be expressed in terms of quadratures involving a fundamental system of solutions of the homogeneous equation.

Proof. Let

$$y_{ij}(x)$$

be a fundamental system of solutions of the homogeneous equation

$$\dot{y}_i = \sum_{j=1}^{n} p_{ij}(x)\, y_j, \qquad i = 1, \ldots, n. \tag{23}$$

Let $u_i(x)$ be n continuously differentiable functions of x to be determined later. Consider the functions

$$\varphi_i(x) = \sum_{\alpha=1}^{n} u_\alpha(x)\, y_{i\alpha}(x), \qquad i = 1, \ldots, n \tag{24}$$

and substitute in Equation (22)

$$\sum_{\alpha=1}^{n} \dot{u}_\alpha y_{i\alpha} + \sum_{\alpha=1}^{n} u_\alpha \dot{y}_{i\alpha} = \sum_{j=1}^{n} p_{ij} \sum_{\alpha=1}^{n} u_\alpha y_{j\alpha} + q_i$$

or

$$\sum_{\alpha} u_\alpha \left(\dot{y}_{i\alpha} - \sum_{j} p_{ij}\, y_{j\alpha}\right) + \sum_{\alpha} \dot{u}_\alpha y_{i\alpha} = q_i.$$

Now the $y_{i\alpha}$ are solutions of Equation (23) and hence the term in parentheses is zero. Hence, if we can determine the $u_\alpha(x)$ such that

$$\sum_{\alpha=1}^{n} \dot{u}_\alpha y_{i\alpha} = q_i \tag{25}$$

our problem will be solved.

Since the y_{ij} form a fundamental system of solutions, their Wronskian

$$W(x) = \begin{vmatrix} y_{11}(x) & \ldots & y_{n1}(x) \\ y_{12}(x) & \ldots & y_{n2}(x) \\ \cdot & \cdot\ \cdot\ \cdot & \cdot \\ y_{1n}(x) & \ldots & y_{nn}(x) \end{vmatrix}$$

is never zero on I. Equation (25) can be considered as a system of linear algebraic equations on the unknowns $\dot{u}_\alpha$ with determinant $W(x) \neq 0$ in I. Hence, by Cramer's rule

$$\dot{u}_k(x) = \frac{1}{W(x)} \begin{vmatrix} y_{11} & y_{21} & \cdots & y_{n1} \\ \cdot & \cdot & \cdot & \cdot \\ y_{1,k-1} & y_{2,k-1} & \cdots & y_{n,k-1} \\ q_1 & q_2 & \cdots & q_n \\ y_{1,k+1} & y_{2,k+1} & \cdots & y_{n,k+1} \\ \cdot & \cdot & \cdot & \cdot \\ y_{1n} & y_{2n} & \cdots & y_{nn} \end{vmatrix}$$

$$= \frac{1}{W(x)} \sum_{j=1}^{n} q_j(x)\, Y_{kj}(x)$$

where Y_{kj} is the cofactor of y_{jk} in $W(x)$.

Therefore

$$u_k(x) = \sum_{j=1}^{n} \int_{x_0}^{x} \frac{1}{W(t)}\, q_j(t)\, Y_{kj}(t)\, dt + C_k \tag{26}$$

where the C_k are constants of integration. From Equation (24)

$$\varphi_i(x) = \sum_{\alpha=1}^{n} y_{i\alpha}(x) \sum_{j=1}^{n} \int_{x_0}^{x} \frac{1}{W(t)}\, q_j(t)\, Y_{\alpha j}(t)\, dt + \sum_{\alpha=1}^{n} C_\alpha\, y_{i\alpha}(x)$$

$$= \sum_{j=1}^{n} \int_{x_0}^{x} \left[\sum_{\alpha=1}^{n} \frac{y_{i\alpha}(x)\, Y_{\alpha j}(t)}{W(t)} \right] q_j(t)\, dt + \sum_{\alpha=1}^{n} C_\alpha\, y_{i\alpha}(x). \tag{27}$$

Now the arbitrary constants of integration, namely the C_α, can be determined by the initial conditions. We have, by hypothesis, $\varphi_i(x_0) = 0$. Applying this criterion to Equations (27) we obtain

$$0 = \varphi_i(x_0) = 0 + \sum_{\alpha=1}^{n} C_\alpha\, y_{i\alpha}(x_0), \qquad i = 1, \ldots, n$$

which is a system of homogeneous linear algebraic equations on the C_α with determinant $W(x_0)$. But $W(x_0) \neq 0$. Hence $C_\alpha = 0$, $\alpha = 1, \ldots, n$ and our solution becomes

$$\varphi_i(x) = \sum_{j=1}^{n} \int_{x_0}^{x} \left[\sum_{\alpha=1}^{n} \frac{y_{i\alpha}(x)\, Y_{\alpha j}(t)}{W(t)} \right] q_j(t)\, dt, \qquad i = 1, \ldots, n. \tag{28}$$

The proof of Theorem 6 is now complete.

4.3 The functions

$$H_{ij}(x, t) = \frac{1}{W(t)} \sum_{\alpha=1}^{n} y_{i\alpha}(x)\, Y_{\alpha j}(t)$$

are called *one-sided Green's functions.* If we interpret Equation (28) in matrix notation,

$$\begin{Vmatrix} \varphi_1(x) \\ \varphi_2(x) \\ \cdots\cdot \\ \varphi_n(x) \end{Vmatrix} = \int_{x_0}^{x} \begin{Vmatrix} H_{11}(x, t) & \ldots & H_{1n}(x, t) \\ H_{21}(x, t) & \ldots & H_{2n}(x, t) \\ \cdot\;\cdot\;\cdot & \cdot\;\cdot\;\cdot & \cdot\;\cdot\;\cdot \\ H_{n1}(x, t) & \ldots & H_{nn}(x, t) \end{Vmatrix} \cdot \begin{Vmatrix} q_1(t) \\ q_2(t) \\ \cdots\cdot \\ q_n(t) \end{Vmatrix} dt, \tag{29}$$

and

$$\begin{Vmatrix} H_{11}(x, t) & \ldots & H_{1n}(x, t) \\ H_{21}(x, t) & \ldots & H_{2n}(x, t) \\ \cdot\;\cdot\;\cdot & \cdot\;\cdot\;\cdot & \cdot\;\cdot\;\cdot \\ H_{n1}(x, t) & \ldots & H_{nn}(x, t) \end{Vmatrix} .$$

is called the one-sided *Green's function matrix.* More explicitly, $H_{ij}(x, t)$ may be written

$$H_{ij}(x,t) = \frac{(-1)^{j-1}}{W(t)} \begin{vmatrix} y_{i1}(x) & y_{i2}(x) & \ldots & y_{in}(x) \\ y_{11}(t) & y_{12}(t) & \ldots & y_{1n}(t) \\ \cdot\;\cdot\;\cdot & \cdot\;\cdot\;\cdot & \cdot\;\cdot\;\cdot & \cdot\;\cdot\;\cdot \\ y_{j-1,1}(t) & y_{j-1,2}(t) & \ldots & y_{j-1,n}(t) \\ y_{j+1,1}(t) & y_{j+1,2}(t) & \ldots & y_{j+1,n}(t) \\ \cdot\;\cdot\;\cdot & \cdot\;\cdot\;\cdot & \cdot\;\cdot\;\cdot & \cdot\;\cdot\;\cdot \\ y_{n1}(t) & y_{n2}(t) & \ldots & y_{nn}(t) \end{vmatrix} . \tag{30}$$

4.4 We see that with the aid of the Green's function we can convert a system of first order linear differential equations (together with boundary conditions) into an integral equation. The integral form is more suitable for many investigations, both theoretical and practical, than the differential equation form. The H_{ij} functions are called *one-sided* Green's functions since the resulting integral equation is of Volterra type. The classical Green's function (yielding a Fredholm integral equation) will not be treated here. It arises most naturally in the theory of linear differential equations with general two point boundary conditions. (Cf. Ince, loc. cit.)

5. The nth order linear differential equation

5.1 One frequently has occasion to consider linear differential equations of the form

$$y^{(n)} + p_1(x)\, y^{(n-1)} + \ldots + p_n(x)\, y = q(x). \tag{31}$$

Equation (31) is called a *linear* differential equation since the unknown function and its derivatives appear in the equation to at most the first degree. In particular, Equation (31) is an nth order equation since the highest derivative appearing is the nth. If $q(x) \equiv 0$, Equation (31) is said to be homogeneous. If $q(x) \not\equiv 0$, Equation (31) is said to be non-homogeneous. In general, the coefficients $p_i(x)$ and $q(x)$ are assumed to be at least continuous on some interval of the x-axis if x is a real variable, or some region in the complex plane if x is a complex variable.

Now Equation (31) can always be written as a first order system of n equations of the type

$$\dot{y}_j = \sum_{k=1}^{n} p_{jk}(x)\, y_k + q_j(x), \qquad j = 1, \ldots, n \tag{32}$$

by introducing the trivial transformations

$$\begin{aligned} \dot{y} &= z_1 \\ \dot{z}_1 &= z_2 \\ &\cdot \quad \cdot \quad \cdot \quad \cdot \quad \cdot \\ \dot{z}_{n-2} &= z_{n-1} \\ \dot{z}_{n-1} &= -\, p_1 z_{n-1} - p_2 z_{n-2} - \ldots - p_n y + q(x). \end{aligned} \tag{33}$$

We see that Equation (32), which is similar to the equations treated earlier, is at least as general as the classical form Equation (31), that is, Equation (31) can always be put in the form of Equation (32).

5.2 However, given a system of type Equation (32), it cannot always be converted into an equation of type Equation (31). Hence Equation (32) is actually more general than Equation (31), and consequently Equation (32) has been discussed in this chapter. Because of the great importance of Equation (31) in its particular form, we shall discuss under what conditions equations given in the form of Equation (32) can be converted to an equation of the form of Equation (31). If we assume that the $p_{ij}(x)$ and $q_i(x)$ are sufficiently differentiable, then in general the problem of solving Equations (32) can be broken down into a problem of solving successively a number of smaller systems; and each of these in turn can be referred to the problem of solving a single equation on one unknown. We shall obtain the result in the form of an algorithm rather than a theorem whose statement would be overly complex because of the many special cases that would have to be considered.

Consider the first equation of Equations (32), viz.:

$$\dot{y}_1 = p_{11}y_1 + p_{12}y_2 + \dots + p_{1n}y_n + q_1 .$$

If $p_{12} = p_{13} = \dots = p_{1n} = 0$, we have an equation on y_1 alone which can be solved and the resulting y_1 inserted into the remaining equations. In this sense then, the system would be resolved into two systems if $p_{1j} = 0$ for $j = 2, \dots, n$. Otherwise we would have a $p_{1j} \neq 0$, and by a re-enumeration of the y_i's, if necessary, we can suppose that $p_{12} \neq 0$. We now make a change of dependent variables in which y_2 is replaced by

$$z_2 = p_{12}y_2 + \dots + p_{1n}y_n$$

and the remaining y_i's are unaffected. One should of course consider in detail the points where $p_{12} = 0$, but we shall not in the present treatment. We now replace the second equation of Equations (32) by an equation obtained by differentiating the above equation and replacing the derivatives of the y_j by their values from Equation (32) and then eliminating y_2.

If the system then does not resolve itself, we can write the first equation in the form

$$\dot{y}_1 = p_{11}y_1 + z_2 + q_1 .$$

We proceed in an analogous way for the second equation. Two cases arise. If the p_{2j}, $j = 3, \dots, n$ are all zero, the system can be resolved into two subsystems, the first of which consists of two equations on two unknowns and must be solved first. Otherwise we may assume that p_{23} is not zero and introduce a second variable,

$$z_3 = p_{23}y_3 + \dots + p_{2n}y_n$$

so that the second equation becomes

$$\dot{z}_2 = p_{12}y_1 + p_{22}z_2 + z_3 + q_2 \ .$$

The continuation of the above process should be clear. We would expect that at some value $r \leq n$ we would obtain a system

$$\begin{aligned}
\dot{y}_1 &= p_{11}y_1 + z_2 + q_1 \\
\dot{z}_2 &= p_{21}y_1 + p_{22}z_2 + z_3 + q_2 \\
&\cdots\cdots\cdots\cdots \\
\dot{z}_r &= p_{r1}y_1 + p_{r2}z_2 + \ldots + p_{rr}z_r + q_r \ .
\end{aligned} \tag{34}$$

The p_{ij} in this and other earlier equations are, of course, not necessarily the same as those that appear in Equation (32).

Now if we differentiate the first of Equations (34) and then replace $\dot{z}_2$ by its value from the second equation, we obtain an equation for $\ddot{y}_1$,

$$\ddot{y}_1 = p^*_{21}y_1 + p^*_{22}z_2 + z_3 + Q_2 \ .$$

If we differentiate and eliminate $\dot{y}_1$, $\dot{z}_2$ and $\dot{z}_3$ we shall obtain an equation on $\dddot{y}_1$ dependent on z_2, z_3, and z_4. We can continue this process until we obtain a system disjoint from the remaining equations:

$$\begin{aligned}
\dot{y}_1 - p_{11}y_1 - q_1 - z_2 &= 0 \\
\ddot{y}_1 - p_{21}y_1 - Q_2 - p_{22}z_2 - z_3 &= 0 \\
\cdots\cdots\cdots\cdots& \\
y_1^{(r-1)} - p_{r-1,1}y_1 - Q_{r-1} - p_{r-1,2}z_2 - \ldots z_r &= 0 \\
y_1^{(r)} - p_{r1}y_1 - Q_r - p_{r2}z_2 - \ldots - p_{rr}z_r &= 0 \ .
\end{aligned} \tag{35}$$

These will yield an rth order differential equation on y_1 if we can eliminate $z_2, \ldots, z_r$ among these r equations. This can readily be done if we consider the system of Equation (35) to be a set of r homogeneous equations on the r quantities $1, z_2, \ldots, z_r$. Since in particular $1 \neq 0$, the determinant of the coefficients of Equation (35), namely

$$\begin{vmatrix} \dot{y}_1 - p_{11}y_1 - q_1 & -1 & 0 & \dots & 0 \\ \ddot{y}_1 - p_{21}y_1 - Q_2 & -p_{22} & -1 & \dots & 0 \\ \dots & \dots & \dots & \dots & \dots \\ y_1^{(r-1)} - p_{r-1,1}y_1 - Q_{r-1} & -p_{r-1,2} & -p_{r-1,3} & \dots & -1 \\ y_1^{(r)} - p_{r1}y_1 - Q_r & -p_{r2} & -p_{r3} & \dots & -p_{rr} \end{vmatrix} \quad (36)$$

must vanish. This is an rth order equation, since in particular the coefficient of $y_1^{(r)}$ is 1. For every solution of Equations (36) we can determine a solution of Equations (35) by taking the corresponding y_1 function and solve the first $r - 1$ equations of Equations (35) for $z_2, \dots, z_r$. Thus it is readily seen that solving Equation (36) is equivalent to solving Equations (35) and the latter in turn is equivalent to solving the first r equations of Equation (32), except possibly for singularities introduced by the need to divide by certain p_{ij}'s.

Thus we have resolved the system of Equation (32) into two systems, the first of which involves r variables, and a remaining system of $n - r$ variables in which r of the original unknowns will now be considered to be known. We may state the following theorem:

Theorem 7. Every function y which is part of a solution of Equation (32) satisfies an rth order differential equation on itself alone where $r \leq n$, and the problem of solving the system of Equation (32) is equivalent to solving a number of smaller systems, each of which is equivalent to solving a differential equation of the form of Equation (31) on one unknown.

5.4 We shall now restate some of our previous theorems in terms of Equation (37),

$$y^{(n)} + p_1(x)\, y^{(n-1)} + \dots + p_n(x)\, y = 0, \quad (37)$$

and also prove some additional results.

Assume that the $p_i(x)$ are continuous in an interval I: [a, b] of the x-axis. Now as remarked earlier, Equation (37) can be converted to a system of first order equations by the trivial transformation:

$$
\begin{aligned}
\dot{z}_1 &= z_2 \\
\dot{z}_2 &= z_3 \\
&\cdots\cdots\cdots\cdots \\
\dot{z}_{n-1} &= z_n \\
\dot{z}_n &= -p_n z_1 - p_{n-1} z_2 - p_{n-2} z_3 - \cdots - p_1 z_n
\end{aligned}
\tag{38}
$$

(for consistency in notation we have set y equal to z_1).

Let

$$\varphi_i(x), \qquad i = 1, \ldots, n$$

be any solution of Equations (38). Then

$$\varphi_1(x)$$

is a solution of Equation (37) and

$$
\begin{aligned}
\varphi_2(x) &= \dot{\varphi}_1(x) \\
\varphi_3(x) &= \dot{\varphi}_2(x) = \ddot{\varphi}_1(x) \\
&\cdots\cdots\cdots\cdots \\
\varphi_{n-1}(x) &= \dot{\varphi}_{n-2}(x) = \ldots = \varphi_1^{(n-2)}(x) \\
\varphi_n(x) &= \dot{\varphi}_{n-1}(x) = \ldots = \varphi_1^{(n-1)}(x).
\end{aligned}
$$

Hence we need only consider $\varphi_1(x)$.

If $\varphi_{ij}(x)$ are a fundamental system of solutions of Equation (38), then

$$\varphi_{11}, \varphi_{12}, \ldots, \varphi_{1n}$$

are solutions of Equation (37). If the $\varphi_{ij}(x)$ are that special fundamental system with the initial conditions

$$\varphi_{ij}(x_0) = \delta_{ij},$$

then

$$\varphi_{11}, \varphi_{12}, \ldots, \varphi_{1n}$$

are solutions of Equation (37) with the properties

$$
\begin{array}{llll}
\varphi_{11}(x_0) = 1 & \dot{\varphi}_{11}(x_0) = 0 & \ldots & \varphi_{11}^{(n-1)}(x_0) = 0 \\
\varphi_{12}(x_0) = 0 & \dot{\varphi}_{12}(x_0) = 1 & \ldots & \varphi_{12}^{(n-1)}(x_0) = 0 \\
\cdot & \cdot & \cdot & \cdot \\
\varphi_{1n}(x_0) = 0 & \dot{\varphi}_{1n}(x_0) = 0 & \ldots & \varphi_{1n}^{(n-1)}(x_0) = 1.
\end{array}
$$

5.5 The Wronskian W(x) of a fundamental system (which is never zero in I) becomes:

$$W(x) = \begin{vmatrix} \varphi_{11}(x) & \dot{\varphi}_{11}(x) & \dots & \varphi_{11}^{(n-1)}(x) \\ \varphi_{12}(x) & \dot{\varphi}_{12}(x) & \dots & \varphi_{12}^{(n-1)}(x) \\ \cdot & \cdot & \cdot & \cdot \\ \varphi_{1n}(x) & \dot{\varphi}_{1n}(x) & \dots & \varphi_{1n}^{(n-1)}(x) \end{vmatrix} . \tag{39}$$

The trace of the matrix on the right hand side of Equations (38) is

$$t(x) = 0 + 0 + \dots + 0 - p_1(x)$$

and hence the Wronskian satisfies the differential equation

$$\dot{W}(x) + p_1(x)\, W(x) = 0.$$

5.6 There is a close connection between linear dependence and solutions of Equation (37) which we shall now investigate. First it will be convenient to drop the subscript "1" on the φ_{1k} functions. Consider, then, the n functions

$$\varphi_1(x),\ \varphi_2(x),\ \dots,\ \varphi_n(x)$$

which is a solution of the differential equation, Equation (37), and which have been derived from a fundamental system of solutions of Equation (38). In this case we know that the Wronskian

$$W(x) = \begin{vmatrix} \varphi_1(x) & \varphi_2(x) & \dots & \varphi_n(x) \\ \varphi_1'(x) & \varphi_2'(x) & \dots & \varphi_n'(x) \\ \cdot & \cdot & \cdot & \cdot \\ \varphi_1^{(n-1)}(x) & \varphi_2^{(n-1)}(x) & \dots & \varphi_n^{(n-1)}(x) \end{vmatrix} \tag{40}$$

is never zero in I. [Note that we have interchanged rows and columns in Equation (39) and used primes instead of dots to indicate first derivatives to obtain the more conventional Equation (40).]

Hence we conclude that since $W(x) \neq 0$ on I, the $\varphi_1(x), \dots, \varphi_n(x)$ functions are *linearly independent* over I. Any set of n functions which satisfy Equation (37) and are linearly independent will be said to form a *fundamental system of solutions* for Equation (37). This is in harmony with our earlier definition of fundamental systems for systems of differential equations, for, if the n functions are linearly independent over I, their Wronskian must of necessity be unequal to zero on I.

5.7 We note that the non-vanishing of the Wronskian is *sufficient* for insuring the linear independence of the $\varphi_1, \ldots, \varphi_n$ functions. Formally stated:

Theorem 8. If $\varphi_1(x), \ldots, \varphi_n(x)$ are n functions which have continuous $(n-1)^{st}$ derivatives in I, then the identical vanishing of their Wronskian is a necessary condition that they be linearly dependent.

However, the identical vanishing of the Wronskian is *not* a sufficient condition that the functions be linearly dependent. Consider the following example.

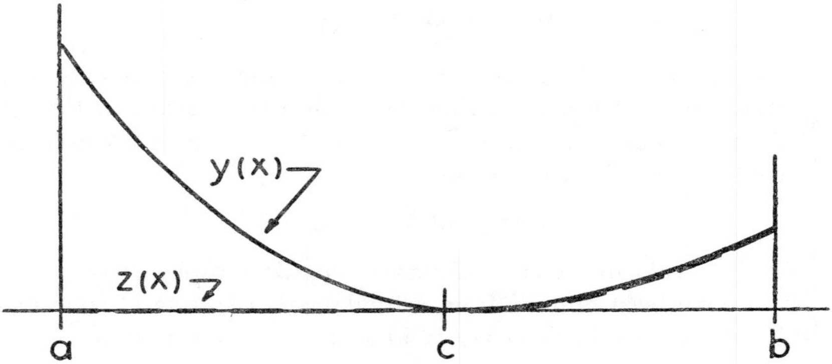

FIGURE 1

Example. Consider the two functions y(x) and z(x). (Cf. Figure 1.) Evidently they can have contact of any order at x = c, and hence z(x) can be assumed to have any finite number of derivatives. y(x) can be taken analytic. Now in [a, c],

$$O \cdot y + Az \equiv 0, \qquad A \neq 0$$

and in [c, b]

$$By - Bz \equiv 0, \qquad B \neq 0.$$

Hence, in either interval,

$$\begin{vmatrix} y(x) & z(x) \\ y'(x) & z'(x) \end{vmatrix} \equiv 0,$$

yet y(x) and z(x) are not linearly dependent except in a piece-wise sense.

In the analytic case, the vanishing of the Wronskian is both a necessary and sufficient condition that the functions be linearly dependent.

5.8 It is trivial to prove that every solution of Equation (37) is a linear combination of a fundamental system (cf. Theorem 5). However, another result, in the spirit of Theorem 8, that is, a converse theorem worded in terms of linear dependence, is the following:

Theorem 9. Let $y(x)$, $y_1(x)$, ..., $y_n(x)$ be $n + 1$ functions which together with their first n derivatives are continuous on I, and such that their Wronskian $W(x)$ vanishes identically on I. Then if the Wronskian $\Delta(x)$ of $y_1(x)$, ..., $y_n(x)$ is never zero on I, there is an identical relationship of the form

$$y(x) = C_1 y_1(x) + \ldots + C_n y_n(x)$$

where C_1, ..., C_n are constants.

Proof. Since $W \equiv 0$, we know that for each value of x there exist quantities u_0, u_1, ..., u_n (not all zero) such that

$$\begin{array}{llll} u_0 y & + u_1 y_1 & + \ldots + u_n y_n & = 0 \\ u_0 y' & + u_1 y_1' & + \ldots + u_n y_n' & = 0 \\ \cdots & \cdots & \cdots & \cdots \\ u_0 y^{(n-1)} & + u_1 y_1^{(n-1)} & + \ldots + u_n y_n^{(n-1)} & = 0 \\ u_0 y^{(n)} & + u_1 y_1^{(n)} & + \ldots + u_n y_n^{(n)} & = 0. \end{array}$$

Consider the first n equations as equations on u_1, ..., u_n. We have, by Cramer's rule

$$u_0 \Delta_r + u_r \Delta = 0, \qquad r = 1, \ldots, n$$

where Δ_r is the determinant Δ in which the rth column has been replaced by $y, y', \ldots, y^{(n-1)}$. If u_0 were zero for any value of x, then since Δ is never zero by hypothesis, all the u_r would be zero — a contradiction. Hence $u_0 \neq 0$. Dividing by u_0 we obtain

$$V_r = -\frac{u_r}{u_0} = \frac{\Delta_r}{\Delta}.$$

(Note that Δ_r and Δ contain only $n - 1$ derivatives of the y, y_i functions while the existence of n derivatives of y, y_i is postulated. Thus the derivative of V_r exists.)

Then

$$
\begin{aligned}
y &= V_1y_1 + V_2y_2 + \dots + V_ny_n \\
y' &= V_1y_1' + V_2y_2' + \dots + V_ny_n' \\
&\cdots\cdots\cdots\cdots\cdots \\
y^{(n)} &= V_1y_1^{(n)} + V_2y_2^{(n)} + \dots + V_ny_n^{(n)}.
\end{aligned}
$$

We shall now show that the V_i are constants, and hence the first of the above equations will yield the theorem.

Differentiate the first n equations and then subtract the rth equation from the $(r - 1)^{st}$, $r = 2, 3, \dots, (n + 1)$. There results,

$$
\begin{aligned}
0 &= V_1'y_1 + V_2'y_2 + \dots + V_n'y_n \\
0 &= V_1'y_1' + V_2'y_2' + \dots + V_n'y_n' \\
&\cdots\cdots\cdots\cdots\cdots \\
0 &= V_1'y_1^{(n-1)} + V_2'y_2^{(n-1)} + \dots + V_n'y_n^{(n-1)}.
\end{aligned}
$$

Since the determinant of the V_i' in the above equations is precisely Δ, and $\Delta \neq 0$, in I,

$$V_i' = 0, \qquad i = 1, 2, \dots, n.$$

It follows therefore that

$$V_i = C_i \quad \text{(a constant)},$$

and our theorem is proved.

5.9 The solution of the non-homogeneous equation

$$y^{(n)} + p_1(x)\, y^{(n-1)} + \dots + p_n(x)\, y = q(x) \tag{41}$$

can be derived from the corresponding theorem for systems of differential equations.

Theorem 10. Let

$$y^{(n)} + p_1(x)\, y^{(n-1)} + \dots + p_n(x)\, y = q(x)$$

be a linear differential equation where $p_i(x)$ and $q(x)$ are continuous in some interval I: [a, b] of the x-axis. Then

$$Y(x) = \int_{x_0}^{x} H(x, t)\, q(t)\, dt$$

is a solution of the non-homogeneous equation with the boundary conditions

$$Y^{(k)}(x_0) = 0, \qquad k = 0, 1, \ldots, n-1, \qquad x_0 \in I,$$

and $H(x, t)$ is the one-sided Green's function

$$H(x,t) = \frac{(-1)^{n-1}}{W(t)} \begin{vmatrix} \varphi_1(x) & \varphi_2(x) & \ldots & \varphi_n(x) \\ \varphi_1(t) & \varphi_2(t) & \ldots & \varphi_n(t) \\ \varphi_1'(t) & \varphi_2'(t) & \ldots & \varphi_n'(t) \\ \cdot & \cdot & \cdot & \cdot \\ \varphi_1^{(n-2)}(t) & \varphi_2^{(n-2)}(t) & \ldots & \varphi_n^{(n-2)}(t) \end{vmatrix}$$

Here $\{\varphi_1, \ldots, \varphi_n\}$ is a fundamental system of solutions of Equation (37) and W is the Wronskian of these functions.

Proof. From Equation (28) (Theorem 6), the solution of the nonhomogeneous equation, Equation (22), is:

$$\varphi_i(x) = \sum_{j=1}^{n} \int_{x_0}^{x} H_{ij}(x, t)\, q_j(t)\, dt. \tag{42}$$

Now if we write Equation (41) as a system of differential equations, it becomes identical with Equations (38) except for the last row which adds $q(x)$,

$$\dot{z}_n = -p_n z_1 - p_{n-1} z_2 - \ldots - p_1 z_n + q.$$

Hence, comparing this with Equation (22) we see that

$$q_j(t) = 0, \qquad j = 1, 2, \ldots, n-1$$
$$q_n(t) = q(t).$$

Equation (42) therefore becomes

$$\varphi_i(x) = \int_{x_0}^{x} H_{in}(x, t)\, q_n(t)\, dt = \int_{x_0}^{x} H_{in}(x, t)\, q(t)\, dt. \tag{43}$$

Also we know that

$$\varphi_2(x) = \varphi_1'(x)$$
$$\varphi_3(x) = \varphi_1''(x)$$
$$\cdots\cdots$$
$$\varphi_n(x) = \varphi_1^{(n-1)}(x).$$

Hence we may confine our attention to $\varphi_1(x)$ which is the desired solution of Equation (41) with the correct boundary conditions, that is,

$$\varphi_1(x_0) = \varphi_1'(x_0) = \ldots = \varphi_1^{(n-1)}(x_0) = 0.$$

Equation (43) then becomes

$$\varphi_1(x) = \int_{x_0}^{x} H_{1n}(x, t)\, q(t)\, dt. \tag{44}$$

From Equation (30)

$$H_{1n}(x, t) = \frac{(-1)^{n-1}}{W(t)} \begin{vmatrix} y_{11}(x) & y_{12}(x) & \dots & y_{1n}(x) \\ y_{11}(t) & y_{12}(t) & \dots & y_{1n}(t) \\ y_{21}(t) & y_{22}(t) & \dots & y_{2n}(t) \\ \cdot & \cdot & \cdot & \cdot \\ y_{n-1,1}(t) & y_{n-1,2}(t) & \dots & y_{n-1,n}(t) \end{vmatrix}. \tag{45}$$

But

$$y_{jk} = y_{1k}^{(j-1)}, \qquad j = 1, \dots, n,$$

and hence Equation (45) becomes

$$H_{1n}(x, t) = \frac{(-1)^{n-1}}{W(t)} \begin{vmatrix} y_{11}(x) & y_{12}(x) & \dots & y_{1n}(x) \\ y_{11}(t) & y_{12}(t) & \dots & y_{1n}(t) \\ y_{11}'(t) & y_{12}'(t) & \dots & y_{1n}'(t) \\ \cdot & \cdot & \cdot & \cdot \\ y_{11}^{(n-2)}(t) & y_{12}^{(n-2)}(t) & \dots & y_{1n}^{(n-2)}(t) \end{vmatrix}.$$

Combining this result with Equation (44) yields Theorem 10.

6. The Jordan theorem

6.1 We have seen that the linear nature of the set of solutions of a system of differential equations

$$\dot{y}_i = \sum_{j=1}^{n} p_{ij}(x)\, y_j \tag{46}$$

permits us to obtain existence theorems for intervals in which the $p_{ij}(x)$ are continuous functions of x. In particular, we have developed a theory concerned with the determinant of values, the Wronskian [cf. Equation (5)]. This theory, in the complex case in which the coefficients $p_{ij}(x)$ are in general analytic, has further implications which we shall now explore. Specifically, the possibilities of the multi-valuedness of the solutions can be specified.

For an arbitrary linear vector space, $\mathfrak{M}$, a linear transformation T is a correspondence, which for each individual vector $W \varepsilon \mathfrak{M}$ uniquely assigns another vector V. This is written $V = T\,W$. Such a correspondence does not depend upon the choice of coordinate systems; but if a choice is made, this axes system determines a matrix A associated with the transformation T. Thus, if $W_1, \ldots, W_n$ is a set of linearly independent unit vectors, we have

$$T\, W_j = \sum_{i=1}^{n} a_{ij}\, W_i \tag{47}$$

and thus the transformation T specifies the matrix $A = \| a_{ij} \|$ for this choice of coordinate systems. If, however, a new coordinate system $W'_1, \ldots, W'_n$ has been chosen, then

$$W'_k = \sum_{h=1}^{n} c_{hk}\, W_h \tag{48}$$

and relative to $W'_1, \ldots, W'_n$ the transformation T will correspond to a matrix A′ where $A' = C^{-1}AC$ and $C = \| c_{ij} \|$. We call A and A′ *similar* matrices. (In general, two square matrices R and S are said to be similar if there exists a non-singular square matrix Q such that $S = Q^{-1}RQ$.) Thus, given T, we will have many similar matrices, A, corresponding to the various ways a coordinate system $W_1, \ldots, W_n$ can be chosen in the linear space $\mathfrak{M}$.

On the other hand, given A, Equation (47) will determine a linear transformation T. Normally this is the way T is specified, although the matrix A is determined not merely by T but by a choice of the coordinate system as well. The inherent properties of T are given, then, by the class of similar matrices $C^{-1}AC$.

Consider, for a given positive integer n, the totality $\mathfrak{N}$ of all $n \times n$ matrices over the complex field. Clearly there exists a subset $\mathfrak{S}$ of $\mathfrak{N}$ with the property that given any matrix A in $\mathfrak{N}$ there exists a matrix $B \in \mathfrak{S}$ such that A is similar to B. If the subset $\mathfrak{S}$ has the property that no two matrices in $\mathfrak{S}$ are similar, then we shall say that the set of matrices $\mathfrak{S}$ forms a *canonical set.* There exist many canonical sets. The form which we shall find most convenient for our investigations is the "Jordan normal form." The result can be considered as a statement on the transformation T.

6.2 We shall now state the Jordan theorem on matrices in a form suitable for our direct applications. The proof of the theorem is available in many treatments of linear problems. For example,

Halmos, P. R. "Finite dimensional vector spaces," Princeton University Press, 1942. Pages 159-169.

Householder, A. S. "Principles of numerical analysis," McGraw-Hill Book Company, Inc., 1953. Pages 30-36.

Mac Duffee, C. C. "An introduction to abstract algebra," John Wiley and Sons, Inc., 1948. Pages 225-242.

Schreier, O., and Sperner, E. "Introduction to modern algebra and matrix theory," Chelsea Publishing Company, 1951. Pages 355-371.

Stoll, R. R. "Linear algebra and matrix theory," McGraw-Hill Book Company, Inc., 1952. Pages 199-209.

Theorem 11. Let T be a linear transformation on an n-dimensional linear vector space $\mathfrak{M}$ and let $\lambda_1, \ldots, \lambda_r$ $(r \leq n)$ be its distinct characteristic roots. Then there exist r sets of vectors

$$\begin{array}{llll} W^{(1,0)}, & W^{(1,1)}, & \ldots, & W^{(1,s_1-1)} \\ W^{(2,0)}, & W^{(2,1)}, & \ldots, & W^{(2,s_2-1)} \\ \cdot & \cdot & \cdot & \cdot \\ W^{(r,0)}, & W^{(r,1)}, & \ldots, & W^{(r,s_r-1)} \end{array}$$

with the following properties:

(i) $s_1 + s_2 + \ldots + s_r = n$.
(ii) The n vectors $W^{(i,j)}$ are linearly independent.
(iii) $T\,W^{(i,0)} = \lambda_i W^{(i,0)}$
(iv) $T\,W^{(i,j)} = \varepsilon_{ij} W^{(i,j-1)} + \lambda_i W^{(i,j)}$ where ε_{ij} is either zero or one.

Because of (ii), the $W^{(i,j)}$ can be used as a basis in the linear vector space $\mathfrak{M}$. If these vectors are used as a basis, then (iii) and (iv) imply that the matrix N corresponding to T in this coordinate system has the form

$$N = \left\| \begin{array}{cccc} B_1 & 0 & \ldots & 0 \\ 0 & B_2 & \ldots & 0 \\ \cdot & \cdot & \cdot & \cdot \\ 0 & 0 & \ldots & B_m \end{array} \right\| \tag{49}$$

where each B_k is a matrix:

$$B_k = \left\| \begin{array}{ccccc} \lambda_i & 0 & \ldots & 0 & 0 \\ 1 & \lambda_i & \ldots & 0 & 0 \\ 0 & 1 & \ldots & 0 & 0 \\ \cdot & \cdot & \cdot & \cdot & \cdot \\ 0 & 0 & \ldots & 1 & \lambda_i \end{array} \right\| \tag{50}$$

and more than one B_k can correspond to the same λ_i, $(m \geq k)$.

If all the characteristic roots are distinct, then N has a purely diagonal representation. Associated with a multiple root there exist more than one normal form. Suppose, for example, $\mathfrak{M}$ is three dimensional and λ is a triple characteristic root. Then the normal form may be any of the following:

$$\left\|\begin{matrix}\lambda & 0 & 0\\ 1 & \lambda & 0\\ 0 & 1 & \lambda\end{matrix}\right\|, \qquad \left\|\begin{matrix}\lambda & 0 & 0\\ 1 & \lambda & 0\\ 0 & 0 & \lambda\end{matrix}\right\|, \qquad \left\|\begin{matrix}\lambda & 0 & 0\\ 0 & \lambda & 0\\ 0 & 0 & \lambda\end{matrix}\right\|.$$

Finally we note that, given a matrix A, we can determine a transformation T by means of Equation (47). For this we have a coordinate system such that the matrix A is in the form of Equation (49). But a change of coordinate system will transform the matrix corresponding to T according to the relation $N = P^{-1}AP$, that is, N is similar to A. Thus every matrix A is similar to a matrix in Jordan normal form. Except possibly for a rearrangement of rows or columns no two Jordan normal forms are similar.

7. The group of monodromy

7.1 Linear differential equations of the form

$$\dot{y}_i = \sum_{j=1}^{n} p_{ij}\, y_j \tag{51}$$

can be conveniently written in the matrix form

$$\dot{y} = R\, y$$

where y is the column vector

$$y = \{y_1, \ldots, y_n\},$$

$\dot{y}$ its derivative and R is the matrix of the coefficients,

$$R = \| p_{ij} \|.$$

When we have a fundamental system of solutions of Equation (51) (cf. Section 3.6 above):

$$\begin{matrix} y_{11}, \ldots, y_{n1} \\ \cdot\ \cdot\ \cdot\ \cdot\ \cdot\ \cdot \\ y_{1n}, \ldots, y_{nn}, \end{matrix}$$

each row can be regarded as a column vector satisfying Equation (51). That is, if we let

$$y^{(i)} = \{y_{1i}, \ldots, y_{ni}\},$$

then

$$\dot{y}^{(i)} \equiv R\, y^{(i)}.$$

(We shall consistently use superscripts for vectors.)

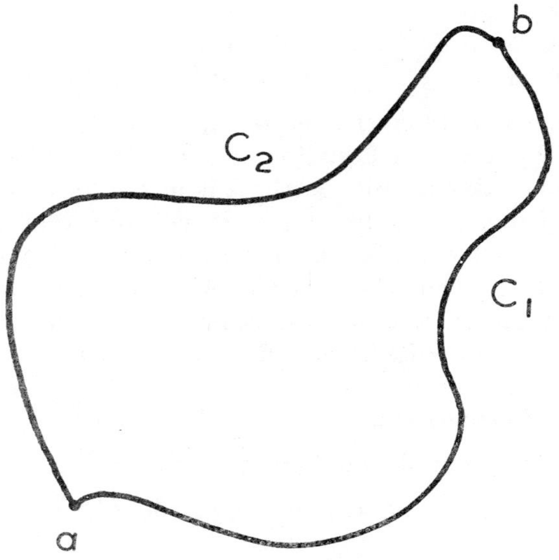

FIGURE 2

7.2 It is trivial to extend Theorem 2 to closed curves consisting of a finite number of simple rectifiable curves. For suppose we have the closed curve made up of C_1 and C_2. (Cf. Figure 2.) Then Theorem 2 applies to C_1 (a to b). At b the solution assumes certain values which may be used as initial conditions in carrying the solution from b to a by another application of Theorem 2. Hence we have carried the solution around the simple closed curve $C_1 + C_2$. Of course, we have tacitly assumed that the $p_{ij}(x)$ coefficients were analytic in some region $\mathfrak{A}$ containing both C_1 and C_2.

7.3 Consider the system of linear differential equations

$$\dot{y}_i = \sum_{j=1}^{n} p_{ij}(x)\, y_j, \qquad i = 1, \ldots, n.$$

Let $\mathfrak{A}$ be the open region in the complex plane on which all the $p_{ij}(x)$ are analytic. (Cf. Figure 3, where the shaded areas and isolated

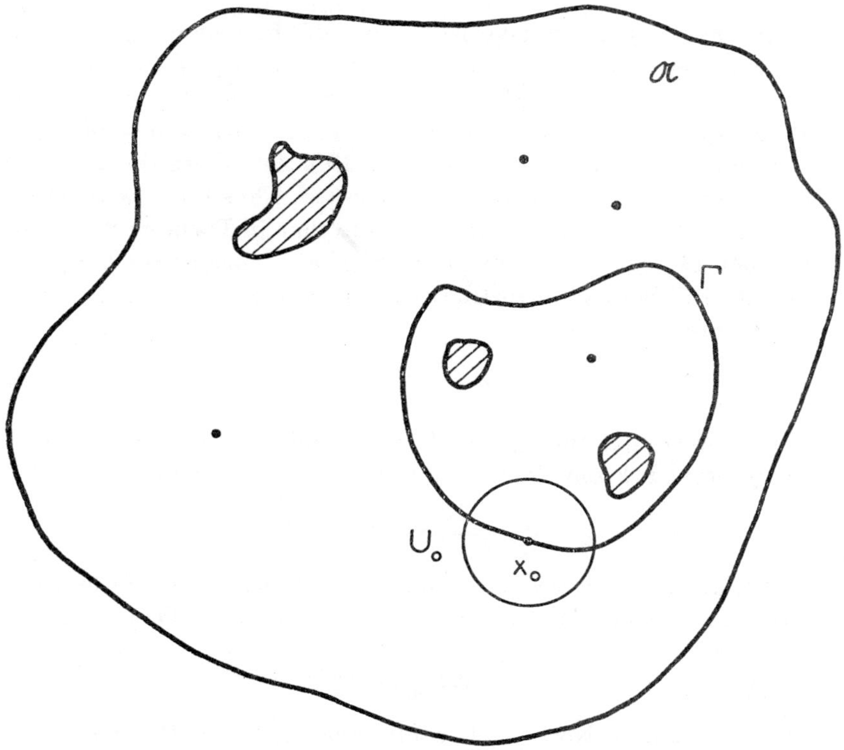

FIGURE 3

points are regions where at least one of the $p_{ij}(x)$ is not analytic.) Let Γ be a simple closed rectifiable curve in $\mathfrak{A}$ and let x_0 be a point on Γ. Now consider any basic system of solutions

$$\begin{aligned} y^{(1)} &= \{y_{11}, \ldots, y_{n1}\} \\ &\cdots\cdots\cdots \\ y^{(n)} &= \{y_{1n}, \ldots, y_{nn}\} \end{aligned} \tag{52}$$

with initial values at x_0, and let U_0 be a neighborhood of x_0 in $\mathfrak{A}$ on which the y_j are analytic. If we make a circuit around Γ, starting and ending at x_0, the $y^{(i)}$ solutions pass into the solutions

$$z^{(1)} = \{z_{11}, \ldots, z_{n1}\}$$
$$\cdots\cdots\cdots \tag{53}$$
$$z^{(n)} = \{z_{1n}, \ldots, z_{nn}\}.$$

In traversing Γ we have passed out of the region U_0, but since we can use an analytic continuation of the y_{ij} at each point of Γ we still have a neighborhood on which these y_{ij} are analytic. When we return to x_0, this analytic continuation may not have the same values as when we started. But the p_{ij} are the same and the differential equations are still satisfied. Thus the new functions z_{ij} are analytic in the region U_0. (Note that in Theorem 2 we have uniqueness because C is a simple arc, that is, has no double points and its end points are distinct.) By Theorem 5, the z's are linear combinations of the y's,

$$z_{ik} = \sum_{j=1}^{n} a_{jk} y_{ij}. \tag{54}$$

We assert that the z_{ik} also form a fundamental system. This is easily proved. From Theorem 3 (for the complex case)

$$W(x) = W(x_0)\, e^{\int_{x_0}^{x} t(\zeta)\, d\zeta}$$

where $t(\zeta)$ is the trace of the matrix R. Since the $y^{(i)}$ form a fundamental system, $W(x_0) \neq 0$. Thus if we circulate $W(x)$ about Γ it becomes

$$W(x_0)\, e^{\int_\Gamma t(\zeta)\, d\zeta}$$

after circulation. Now $t(\zeta)$ is continuous on Γ and Γ is rectifiable. Thus $\int_\Gamma t(\zeta)\, d\zeta$ is finite and hence

$$W(x_0)\, e^{\int_\Gamma t(\zeta)\, d\zeta} \neq 0.$$

We conclude, therefore, that the z's form a fundamental system.

7.4 Suppose now that $\{y_1, \ldots, y_n\}$ is an arbitrary vector solution of Equations (51). Then we know that this vector can be expressed as a linear combination of $y^{(i)}$ vectors:

$$y_i = \sum_{k=1}^{n} \alpha_k y_{ik}, \qquad i = 1, \ldots, n.$$

When we make a circuit around Γ with this arbitrary solution the solution passes into the new solution z_i,

$$z_i = \sum_{k=1}^{n} \alpha_k z_{ik} = \sum_k \alpha_k \sum_j a_{jk} y_{ij} = \sum_j \left(\sum_k \alpha_k a_{jk}\right) y_{ij}.$$

Let

$$\beta_j = \sum_k \alpha_k a_{jk}.$$

Then

$$z_i = \sum_j \beta_j y_{ij}.$$

So if the original arbitrary solution was given by the vector $\{\alpha_1, \ldots, \alpha_n\}$ (that is, these constants times the fundamental system $y^{(i)}$,

$$y = \sum_k \alpha_k y^{(k)}$$

where y is the column vector $\{y_1, \ldots, y_n\}$) then the new solution is given by the vector $\{\beta_1, \ldots, \beta_n\}$. The relation between the α's and β's is

$$\beta = A\,\alpha$$

where A is the matrix

$$A = \| a_{ij} \| = \left\| \begin{matrix} a_{11} & \ldots & a_{1n} \\ \cdot & \cdot\ \cdot\ \cdot & \cdot \\ a_{n1} & \ldots & a_{nn} \end{matrix} \right\|,$$

β is the column vector $\{\beta_1, \ldots, \beta_n\}$ and α is the column vector $\{\alpha_1, \ldots, \alpha_n\}$. Hence any solution "$\alpha$" is transformed into a solution "β" by the non-singular transformation A. ($|A| \neq 0$ since from Equation (54) we may write $Z = A'Y$ where Z is the matrix of the z solutions

$$Z = \left\| \begin{matrix} z_{11} & \ldots & z_{n1} \\ \cdot & \cdot\ \cdot\ \cdot & \cdot \\ z_{1n} & \ldots & z_{nn} \end{matrix} \right\|,$$

Y is similarly defined and A′ is the transpose of A. Now $|Y| \neq 0$, $|Z| \neq 0$ since they are fundamental systems. Hence from $|Z| = |A'| \cdot |Y|$ we conclude $|A'| \neq 0$ and therefore $|A| \neq 0$.) Thus corresponding to A we have a non-singular transformation T on the linear set of solutions.

7.5 If Γ_1 is any other contour starting and ending at x_0, and if any solution $\varphi = \{\varphi_1, \ldots, \varphi_n\}$ is carried around Γ_1 in $\mathfrak{A}$, then

φ will be taken into a new solution ψ which can be related to φ by a non-singular transformation T_1,

$$\psi = T_1\varphi.$$

Furthermore, slight variations in Γ_1 will not affect T_1, that is, if one contour can be obtained from another by a continuous deformation which does not pass out of the region of analyticity $\mathfrak{A}$ of the p_{ij}, then the transformations obtained from these two contours will be identical since the solutions will be analytic on these contours and between them. We shall call such contours "equivalent." Equivalent contours surround the same singularities of the p_{ij}. If we traverse Γ and Γ_1 or take a path equivalent to such a traversal we obtain a $T_2 = T_1T$.

If a contour can be continuously shrunk to a point while always remaining within the region of analyticity of the p_{ij}, that is, in $\mathfrak{A}$, then the corresponding transformation is the identity transformation. If a contour with transformation T is traversed in the clockwise direction, the corresponding transformation is the inverse of T.

One can readily see that the totality of the transformations T form a *group*. This is the *monodromic group* of the system of linear transformations, Equation (51) (cf. Ince, loc. cit., p. 389). It may seem at first that this is a group with respect to the point x_0, but if $\mathfrak{A}$ is connected one can readily show that the groups associated with different points are isomorphic.

7.6 The problem we wish to investigate now is: What is the nature of the multi-valuedness that appears in the solutions?

Suppose, as before, that we have an arbitrary solution given by the vector $\{\alpha_1, \ldots, \alpha_n\}$ which passes into the solution $\{\beta_1, \ldots, \beta_n\}$ where $\beta = A\alpha$. Suppose further that $\{\alpha_1, \ldots, \alpha_n\}$ is a characteristic vector of A. Then

$$A\,\alpha = \lambda\,\alpha$$

where λ is a characteristic root (in general a complex number). Therefore

$$\beta = A\,\alpha = \lambda\,\alpha$$

and hence

$$\beta_i = \lambda\,\alpha_i.$$

Since, further

$$z_i = \sum_j \beta_j y_{ij},$$

it follows that

$$z_i = \sum_j \lambda \alpha_j y_{ij} = \lambda \sum_j \alpha_j y_{ij} = \lambda y_i, \qquad i = 1, \ldots, n.$$

We shall now suppose that a is an isolated singular point interior to Γ, for some of the p_{ij} functions. That is, $x = a$ is an isolated point at which at least one p_{ij} is not analytic. We shall assume that a is the only singular point within Γ.

Consider now the function $(x\text{-}a)^r$ (which is analytic in any circle not containing a) where r is, in general, a complex number. We may write this as

$$(x\text{-}a)^r = e^{r \log (x-a)}.$$

If we make one circuit of log (x-a) around Γ, we obtain

$$\log (x' - a) = \log (x - a) + 2\pi i$$

where x' is the same point as x after a circulation [on a different sheet of the Riemann surface for $\log (x - a)$]. Therefore,

$$(x' - a)^r = e^{r \log (x'-a)} = e^{r \log (x-a) + r2\pi i} = (x - a)^r e^{2\pi ir}.$$

Thus we have shown that on a circuit around Γ, where Γ contains the point a, $(x - a)^r$ changes by a factor $e^{2\pi ir}$ to $(x' - a)^r$.

Since all the characteristic roots of A are unequal to zero (since $|A| \neq 0$), we may consider $\log \lambda$ where λ is a characteristic root of A. Let

$$r = \frac{\log \lambda}{2\pi i}.$$

Then if we circulate $(x - a)^r$ around Γ we obtain

$$(x' - a)^r = e^{2\pi i \frac{\log \lambda}{2\pi i}} (x - a)^r = e^{\log \lambda} (x - a)^r = \lambda (x - a)^r.$$

That is, in the case where $r = (\log \lambda)/2\pi i$ we have shown that on a circuit around Γ, $(x - a)^r$ changes by the factor λ.

Consider now the vector $y_1, \ldots, y_n$ corresponding to α and the functions

$$\frac{y_i}{(x-a)^r}$$

where

$$r = \frac{\log \lambda}{2\pi i}.$$

In circulating once around Γ we obtain

$$\frac{y_i'}{(x'-a)^r} = \frac{z_i}{\lambda(x-a)^r} = \frac{\lambda y_i}{\lambda(x-a)^r} = \frac{y_i}{(x-a)^r}.$$

That is, we obtain the function itself back. So we have proved

Theorem 12. If $y_1, \ldots, y_n$ corresponds to $\alpha_1, \ldots, \alpha_n$ where $\alpha_1, \ldots, \alpha_n$ is a characteristic vector of A, then

$$y_i = (x-a)^r \varphi_i(x)$$

where $\varphi_i(x)$ is analytic in a ring around a and $r = (\log \lambda)/2\pi i$ where λ is the characteristic root corresponding to α.

7.7 We recall from the Jordan theorem that the n characteristic vectors $W^{(i,j)}$ were linearly independent and hence formed a basis. That is, any vector can be written as a linear combination of these characteristic vectors. Hence if we can show how the solutions which correspond to characteristic vectors behave, we shall have determined the behavior of solutions in the general case.

Before using the full power of Theorem 11 we consider the special case where T has n distinct characteristic roots

$$\lambda_1, \ldots, \lambda_n$$

and corresponding characteristic vectors

$$\begin{aligned} \alpha^{(1)} &= \{\alpha_{11}, \ldots, \alpha_{n1}\} \\ &\cdots\cdots\cdots \\ \alpha^{(n)} &= \{\alpha_{1n}, \ldots, \alpha_{nn}\}. \end{aligned}$$

From (ii) of Theorem 11 we know that the $\alpha^{(i)}$ are linearly independent and hence their matrix $\|\alpha_{ij}\|$ is non-singular.

Let

$$w_{1k}, \ldots, w_{nk}$$

be the solution corresponding to the k*th* characteristic vector. If we circulate w_{1k} once around Γ, w_{1k} passes into $\lambda_k w_{1k}$. Hence

$$w_{1k} = (x-a)^{r_k} \varphi_{1k}(x)$$

where $\varphi_{1k}(x)$ is a function analytic in a ring around a and $r_k = (\log \lambda_k)/2\pi i$.

Now, since the $\| \alpha_{ij} \|$ matrix is non-singular, the w_{ij} constitute a fundamental system of solutions. Hence any solution can be expressed as a linear combination of these. So for i fixed, an arbitrary solution y_i is

$$y_i = \sum_j c_j w_{ij} = \sum_j c_j (x - a)^{r_j} \varphi_{ij}(x)$$

where the φ_{ij} are analytic in a ring around a.

7.8 There remains but to consider the case where T does not have distinct characteristic roots, or more precisely where the matrix N of Equation (49) corresponding to T does not have pure diagonal Jordan normal form. We shall consider first the case in which N is a 2×2 matrix with a double root and normal form

$$\left\| \begin{matrix} \lambda & 0 \\ 1 & \lambda \end{matrix} \right\| . \qquad (55)$$

This case will illustrate the ideas involved.

By Theorem 11, there exist two vectors w(x) and v(x) such that

$$T\, w(x) = \lambda\, w(x) \qquad (56)$$

$$T\, v(x) = w(x) + \lambda\, v(x). \qquad (57)$$

For Equation (56) we prove as above that

$$w(x) = (x - a)^r \varphi(x)$$

where $r = (\log \lambda)/2\pi i$ and $\varphi(x)$ is a function analytic in a ring around a.

Now consider Equation (57). If v(x) makes a circuit around Γ, it goes over into T v(x),

$$T\, v(x) = w(x) + \lambda\, v(x).$$

Also, from Equation (56)

$$T\, w(x) = \lambda\, w(x).$$

Therefore, dividing these two equations

$$\frac{T\, v(x)}{T\, w(x)} = T\left[\frac{v(x)}{w(x)}\right] = \frac{1}{\lambda} + \frac{v(x)}{w(x)}.$$

(It may be necessary to deform Γ slightly so as to avoid the zeros of $w(x)$.) Let

$$q(x) = \frac{v(x)}{w(x)}\cdot$$

Then

$$\frac{Tv}{Tw} = T\left(\frac{v}{w}\right) = Tq = \frac{1}{\lambda} + q. \tag{58}$$

If we circulate $\frac{1}{2\pi i\lambda}\log(x-a)$ around Γ we obtain

$$\frac{1}{2\pi i\lambda}\log(x'-a) = \frac{1}{2\pi i\lambda}[\log(x-a)+2\pi i] = \frac{1}{2\pi i\lambda}\log(x-a)+\frac{1}{\lambda}.$$

Hence, any solution of Equation (58) is of the form

$$q = \frac{1}{2\pi i\lambda}\log(x-a) + \psi(x)$$

where $\psi(x)$ is single valued and analytic in a ring around a. But

$$q(x) = \frac{v(x)}{w(x)}.$$

Hence

$$v(x) = \frac{w(x)}{2\pi i\lambda}\log(x-a) + w(x)\,\psi(x) = (x-a)^r[\varphi^*(x)\log(x-a)+\psi^*(x)]$$

where φ^* and ψ^* are single valued and analytic in a ring around a.

7.9 For a matrix of arbitrary normal form the above process may be generalized to obtain the form of the solutions for the characteristic vectors, and hence of an arbitrary solution which is a linear combination of these.

Consider, for instance, the case of a third order box for which the corresponding Jordan normal form is

$$\begin{Vmatrix} \lambda & 0 & 0 \\ 1 & \lambda & 0 \\ 0 & 1 & \lambda \end{Vmatrix}.$$

Here

$$\begin{aligned} T\,w(x) &= \lambda\,w(x) \\ T\,v(x) &= w(x) + \lambda\,v(x) \\ T\,u(x) &= v(x) + \lambda\,u(x). \end{aligned}$$

The arguments given above for w and v still apply and thus we still have the form given above for q(x). However, to determine the form of u, we introduce

$$p(x) = \frac{u(x)}{w(x)}.$$

Then in a manner similar to that in which Equation (58) was obtained we arrive at

$$\frac{Tu}{Tw} = Tp = \frac{1}{\lambda}\frac{v}{w} + p = \frac{1}{\lambda}q + p$$
$$= \frac{1}{2\pi i\lambda^2}\log (x - a) + \frac{1}{\lambda}\psi(x) + p(x).$$

Now we can solve this equation provided we can solve the equations:

$$T\,\pi_1(x) = \frac{1}{2\pi i\lambda^2}\log (x - a) + \pi_1$$
$$T\,\pi_2(x) = \frac{1}{\lambda}\psi(x) + \pi_2.$$

For by addition we see that π_1 plus π_2 is a solution of the equation for p and hence any other solution differs from this only by a single valued function. But

$$\pi_1 = -\frac{1}{8\pi^2\lambda^2}[\log (x - a)]^2 - \frac{1}{4\pi i\lambda^2}\log (x - a)$$
$$\pi_2 = \frac{\psi(x)}{2\pi i\lambda}\log (x - a)$$

are solutions to these equations. (By a direct substitution one can always find a solution to

$$T\,\omega = A\,[\log (x - a)]^k + \omega$$

in the form of a polynomial in log (x — a) of degree k + 1.)

Thus p is of the form

$$\pi_1 + \pi_2 + \chi(x)$$

where $\chi(x)$ is single valued and analytic in a ring around a. Consequently we can conclude that u is of the form

$$u(x) = (x - a)^r\,[\varphi^*(x) \log^2(x - a) + \psi^*(x) \log (x - a) + \chi^*(x)].$$

It is clear how this generalizes to higher order boxes and of course one can readily obtain the result for the case of a number of boxes.

7.10 The results thus obtained are highly significant in connection with the problem of the solution of differential equations by a series of powers of x — a. The functions φ, ψ and χ, etc., obtained above are analytic in a ring around a and hence have Laurent expansions such as

$$\varphi(x) = \sum_{n=-\infty}^{\infty} A_n (x - a)^n.$$

Thus when the transformation T has n distinct characteristic roots we know that there are solutions in the form

$$\sum_{n=-\infty}^{\infty} A_n (x - a)^{n+r_k}$$

where of course $r_k = (\log \lambda_k)/2\pi i$. In the case of equal roots we may find terms with a factor log (x — a). However, these logarithmic solutions have a rather specialized form of which the equation given in Section 7.8 for v(x) is typical.

The objective of the above investigation is to specify the multi-valued character of the solutions. The functions φ, ψ, χ, etc., may well have essential singularities at a. The conditions of Fuchs (cf. Ince, loc. cit., p. 365) are necessary and sufficient that the solutions do not have an essential singularity at a. For the second order equation

$$y'' + p_1(x)\, y' + p_2(x)\, y = 0$$

this theorem states that a necessary and sufficient condition that the solutions of this equation do not have an essential singularity at a is that $(x - a)\, p_1(x)$ and $(x - a)^2\, p_2(x)$ be analytic in a neighborhood of a. These conditions are precisely those which permit one to obtain successively the coefficients of a power series which extends in one direction only, that is, which does not have arbitrarily high negative powers (cf. Ince, loc. cit., p. 396).

8. Linear differential equations with constant coefficients

8.1 The remaining subject we wish to consider is that of linear differential equations with *constant coefficients*. This type of equation is of great practical interest since it is the only general class of equations for which an explicit solution can be found. Consider the system

$$\dot{y}_i = \sum_{j=1}^{n} a_{ij} y_j, \qquad i = 1, \ldots, n \tag{59}$$

where the a_{ij} are *constants* (real or complex). We may write Equation (59) in matrix form

$$\dot{y} = A y \tag{60}$$

where y is the column vector $\{y_1, \ldots, y_n\}$, $\dot{y}$ its derivative and A is the constant matrix $\| a_{ij} \|$. From Theorem 11 we know that there exists a non-singular matrix P such that

$$N = P^{-1}AP$$

is in Jordan normal form.

We may multiply Equation (60) on the left by P^{-1} and we obtain

$$P^{-1} \dot{y} = P^{-1} A y.$$

But

$$P^{-1}Ay = P^{-1}A(PP^{-1})y = (P^{-1}AP)(P^{-1}y). \tag{61}$$

If we let

$$z = P^{-1} y$$

then

$$\dot{z} = P^{-1} \dot{y}$$

and Equation (61) becomes

$$\dot{z} = N z$$

where N is in Jordan normal form.

8.2 We know that N can be written as

$$N = \left\| \begin{matrix} B_1 & 0 & \ldots & 0 \\ 0 & B_2 & \ldots & 0 \\ \cdot & \cdot & \cdot & \cdot \\ 0 & 0 & \ldots & B_m \end{matrix} \right\|$$

where B_k is the matrix

$$B_k = \left\| \begin{matrix} \lambda_i & 0 & \ldots & 0 & 0 \\ 1 & \lambda_i & \ldots & 0 & 0 \\ \cdot & \cdot & \cdot & \cdot & \cdot \\ 0 & 0 & \ldots & 1 & \lambda_i \end{matrix} \right\|$$

and the λ_i are the characteristic roots of A.

Consider a typical B block

$$\begin{Vmatrix} \lambda & 0 & \dots & 0 & 0 \\ 1 & \lambda & \dots & 0 & 0 \\ . & . & . & . & . \\ 0 & 0 & \dots & 1 & \lambda \end{Vmatrix}$$

and the equation

$$\begin{Vmatrix} \dot{z}_1 \\ \dot{z}_2 \\ \vdots \\ \vdots \\ \dot{z}_r \end{Vmatrix} = \begin{Vmatrix} \lambda & 0 & \dots & 0 & 0 \\ 1 & \lambda & \dots & 0 & 0 \\ . & . & . & . & . \\ . & . & . & . & . \\ 0 & 0 & \dots & 1 & \lambda \end{Vmatrix} \cdot \begin{Vmatrix} z_1 \\ z_2 \\ \vdots \\ \vdots \\ z_r \end{Vmatrix}. \tag{62}$$

Now Equation (62) is equivalent to the r scalar equations

$$\begin{aligned} \dot{z}_1 &= \lambda z_1 \\ \dot{z}_2 &= z_1 + \lambda z_2 \\ \dot{z}_3 &= z_2 + \lambda z_3 \\ &. \quad . \quad . \quad . \quad . \quad . \\ \dot{z}_r &= z_{r-1} + \lambda z_r. \end{aligned} \tag{63}$$

Suppose we wish to find the solution of the above system which at $x = x_0$ assumes the values

$$z_1(x_0) = 1, \qquad z_2(x_0) = 0, \qquad \dots, \qquad z_r(x_0) = 0.$$

Clearly, the solution of

$$\dot{z}_1 = \lambda z_1, \qquad z_1(x_0) = 1$$

is

$$z_1(x) = e^{\lambda(x-x_0)}.$$

The solution of the non-homogeneous system

$$\dot{z}_2 = \lambda z_2 + z_1, \qquad z_2(x_0) = 0$$

is evidently

$$z_2(x) = (x - x_0)\, e^{\lambda(x-x_0)}.$$

In general we easily determine that

$$z_k(x) = \frac{(x - x_0)^{k-1}}{(k-1)!} e^{\lambda(x-x_0)}, \qquad k = 1, \dots, r.$$

Hence the vector solution $\varphi^{(1)}(x)$ with the initial conditions $(1, 0, \dots, 0)$ is

$$\varphi^{(1)}(x) = \left\{ e^{\lambda(x-x_0)}, \quad (x - x_0)\, e^{\lambda(x-x_0)}, \ldots, \frac{(x - x_0)^{r-1}}{(r-1)!} e^{\lambda(x-x_0)} \right\}.$$

Now consider the solution $\varphi^{(2)}(x)$ with the initial conditions $(0, 1, 0, \ldots, 0)$. The solution of

$$\dot{z}_1 = \lambda z_1, \qquad z_1(x_0) = 0$$

is

$$z_1(x) \equiv 0.$$

The solution of

$$\dot{z}_2 = \lambda z_2 + z_1, \qquad z_2(x_0) = 1$$

is

$$z_2(x) = e^{\lambda(x-x_0)}$$

and so on. We obtain

$$\varphi^{(2)}(x) = \left\{ 0, e^{\lambda(x-x_0)}, \ldots, \frac{(x - x_0)^{r-2}}{(r-2)!} e^{\lambda(x-x_0)} \right\} .$$

If we continue the process we find that the solution $\varphi^{(r)}(x)$ with the initial conditions $(0, 0, \ldots, 1)$ is

$$\varphi^{(r)}(x) = \left\{ 0, 0, \ldots, e^{\lambda(x-x_0)} \right\}.$$

8.3 So we may state our results as follows:

Theorem 13. If $\dot{z} = Nz$ where N is in Jordan normal form, then corresponding to each block B_k we have r solutions

$$\begin{aligned}
\varphi^{(1)}(x) &= \left\{ e^{\lambda(x-x_0)}, (x - x_0)\, e^{\lambda(x-x_0)}, \ldots, \frac{(x - x_0)^{r-1}}{(r-1)!} e^{\lambda(x-x_0)} \right\} \\
\varphi^{(2)}(x) &= \left\{ \quad 0 \quad , \quad e^{\lambda(x-x_0)} \quad , \ldots, \frac{(x - x_0)^{r-2}}{(r-2)!} e^{\lambda(x-x_0)} \right\} \\
&\cdots\cdots\cdots\cdots\cdots\cdots \\
\varphi^{(r)}(x) &= \left\{ \quad 0 \quad , \quad 0 \quad , \ldots, \quad e^{\lambda(x-x_0)} \quad \right\}
\end{aligned}$$

with the corresponding initial values:

$$\begin{aligned}
\varphi^{(1)}(x_0) &= \{1, 0, \ldots, 0\} \\
\varphi^{(2)}(x_0) &= \{0, 1, \ldots, 0\} \\
&\cdots\cdots\cdots \\
\varphi^{(r)}(x_0) &= \{0, 0, \ldots, 1\}
\end{aligned}$$

at $x = x_0$.

Index

ERRATA

EXISTENCE THEOREMS FOR ORDINARY DIFFERENTIAL EQUATIONS

Title page. Fourth line should read:

Professor of Mathematics, Duke University

Page 24. Last line before footnote should read:

$$| z^*_{ri} - z_{r,0} | < b.$$

Page 34. Lines 2 and 3 should read:

$$F_i(y'_{d_1}, y_{d_1}, y_{d_1-1}, \ldots, y_1, \quad y'_{d_1+d_2}, y_{d_1+d_2}, \ldots, y_{d_1+1}, \ldots,$$
$$y'_{d_1+\cdots+d_r}, y_{d_1+\cdots+d_r}, \ldots, y_{d_1+\cdots+d_{r-1}+1}, x) = 0, \quad i = 1, \ldots, r$$

Page 37. Line 3 should read:

$\mathfrak{A}$ of two-dimensional euclidean space, then f is said to satisfy a

and lines 3 and 4 in paragraph *1.4* should read:

convex open region $\mathfrak{A}$. We know that f(y, x) is continuous in y alone if, given an $\varepsilon > 0$ and two points $(y_1, x), (y_2, x) \varepsilon \mathfrak{A}$, there

Page 38. Line 3 in *Theorem 1. Hypothesis,* should read:

open region $\mathfrak{A}$ of two-dimensional euclidean space. Let f(y, x)

Page 42. Line 5 should read:

tinuous on a convex open region $\mathfrak{A}$ of (n + 1)-dimensional euclidean

Page 44. Two lines before caption **4. Uniqueness** . . . should read:

$$| y_j - z_j | \leq \int_{x'}^{x} | f_j(y_1, \ldots, y_n, t) - f_j(z_1, \ldots, z_n, t) | \, dt, \quad j = 1, \ldots n.$$

Page 45. Line 11 should read:

continuous on a convex open region $\mathfrak{A}$ of (2n + 1)-dimensional

Page 50. Line 12 should read:

dition in this region. Then for every point (y_0, x_0) of $\mathfrak{A}$ we can find

Page 58. Line 2 in *Conclusion.* should read:

with the initial iterant $y_j^{(0)}(x) = y'_j$, then the $y_j^{(k)}(x)$ are defined

Page 82. Line 12 should read:

$$| f_y(y + \alpha z, x) z - f_y(y', x')z' | < \varepsilon.$$

Page 90. Line 5 should read:

show that if $(x, z'_1, \ldots, z'_n, \Delta y^*_1)$ and $(x, z''_1, \ldots, z''_n, \Delta y^*_1)$ are two

Line 20 should read:

being a point intermediate between z'_k and z''_k. The $\partial f_i / \partial y_k$ are

Page 102. Line 18 should read:

(ii) Let $\dfrac{\partial^s f_1}{\partial x^{s_0} \partial y_1^{s_1} \ldots \partial y_n^{s_n} \partial \lambda_1^{s_{n+1}} \ldots \partial \lambda_r^{s_{n+r}}}$ exist and be jointly continu-

ROBERT E. KRIEGER PUBLISHING COMPANY, INC.
POST OFFICE BOX 542, HUNTINGTON, N. Y. 11743